THE SCIENCE LOVER'S GUIDE TO LONDON

THE SCIENCE LOVER'S GUIDE TO LONDON

RACHAEL ROWE

WHITE OWL
AN IMPRINT OF PEN & SWORD BOOKS LTD.
YORKSHIRE – PHILADELPHIA

First published in Great Britain in 2024 by
White Owl
An imprint of
Pen & Sword Books Ltd.
Yorkshire - Philadelphia

ISBN 978 1 39906 362 3

A CIP catalogue record for this book is available from the British Library.

Printed and bound by Printworks Global Ltd, London/Hong Kong.
Design: SJmagic DESIGN SERVICES, India.

Pen & Sword Books Ltd. incorporates the imprints of Pen & Sword Books: After the Battle, Archaeology, Atlas, Aviation, Battleground, Discovery, Family History, History, Maritime, Military, Politics, Select, Transport, True Crime, Fiction, Frontline Books, Leo Cooper, Praetorian Press, Seaforth Publishing, Wharncliffe and White Owl.

For a complete list of Pen & Sword titles please contact

PEN & SWORD BOOKS LIMITED
George House, Beevor Street, Off Pontefract Road, Hoyle Mill, Barnsley, South Yorkshire, England, S71 1HN.
E-mail: enquiries@pen-and-sword.co.uk
Website: www.pen-and-sword.co.uk

or

PEN AND SWORD BOOKS
1950 Lawrence Rd, Havertown, PA 19083, USA
E-mail: uspen-and-sword@casematepublishers.com
website: www.penandswordbooks.com

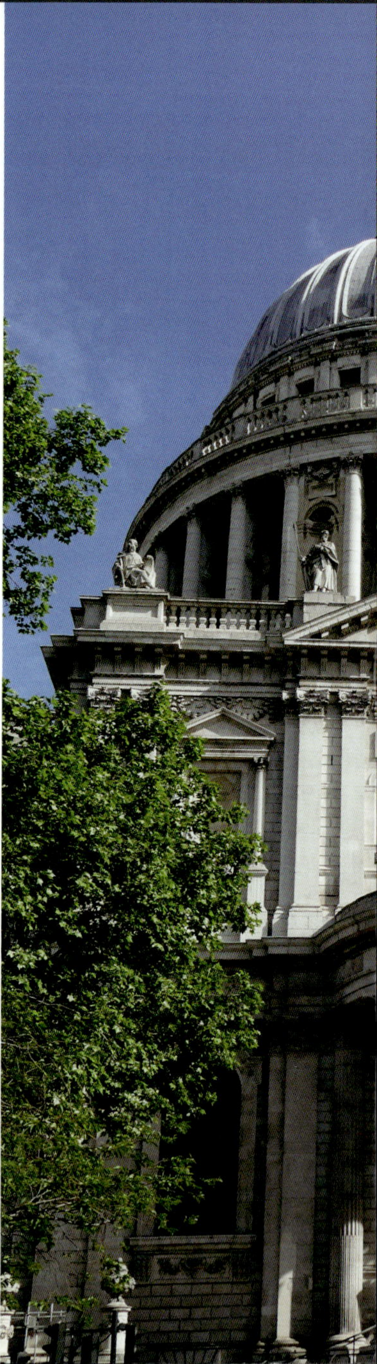

CONTENTS

FOREWORD

'Science and everyday life cannot and should not be separated.'
– Rosalind Franklin

Rosalind Franklin, the scientist who was pivotal in the discovery of DNA, viewed science as an integral part of life. Wherever you look in London, science is all around, from artificial intelligence used in the public transport system, to the engineering behind Tower Bridge and high-rise buildings. Although science takes place all over Britain and across the world, London is a hub for museums, specialist collections, scientific societies, and research.

There have been scientific developments in London for centuries. Much of the Roman heritage has been paved over, but it is underground where ancient engineering techniques to create baths and other structures are found. St Bartholomew's Hospital has been at the forefront of medical developments for 900 years and is Britain's oldest hospital.

As the Age of Enlightenment dawned in London, scientific observation and experiments created new ways of understanding the world. Science was viewed as a foundation for success in industry and commerce. Opportunities to innovate and build better came with astronomy at the Royal Observatory in Greenwich, the rebuilding of St Paul's Cathedral, and the theories developed by Sir Isaac Newton.

Although many people came to London to live and study, some were born within the metropolis. For example, Michael Faraday was an extraordinary scientist whose electricity generation and transmission discoveries have helped shape the modern world.

With an increase in population came the need to develop infrastructure and engineering. The remarkable bascule structure of Tower Bridge, the work of Mark Brunel in engineering the Thames Tunnel, and the safety testing by David Kirkaldy all helped shape London. The precision of weights and loads by London's dock workers is a prime example of how science is integral to everyday life.

However, with a growing city came the need to control disease and waste. Dr John Snow's studies in epidemiology saved thousands of lives when he identified the source of a cholera outbreak in 1854. And the need to manage sewage after The Great Stink saw the pioneering work of Joseph Bazalgette transform the Thames and waste disposal in London with the

Tower Bridge, London. (Sekau67, Pixabay)

creation of the Embankment and iconic pumping stations.

Today, London continues to transform and change, with new scientific discoveries occurring all the time. The skyline features iconic buildings with engineering genius at the centre. Sustainability is prominent with botanical research at Kew Gardens and innovative ways of using green spaces. There are museums, free lectures, and experiences all over the city where anyone can be inspired by science and its ability to transform the world. This is the first colour guidebook to science in London and includes a wealth of interesting places to discover, and find out more. Science is indeed integral to the way we live and view a city.

THE CITY OF LONDON

Renowned for its rich history and prominent finance district, the square mile of the City of London is one of the oldest places in the capital. Narrow streets full of medieval history contrast with some of the most dynamic architectural designs in the United Kingdom. Scientific innovation and heritage have been at the heart of the City of London for centuries, dating from the Romans. After the Great Fire of London destroyed the City in 1666, mathematicians and scientists at the Royal Society stepped forward to redesign buildings, including St Paul's Cathedral. Where high-rise banks and

St Paul's Cathedral. (Author)

commercial businesses now operate, scientific instrument makers once crafted tools to support developments in navigation and time. The area around Smithfield was once notorious for its execution site and medieval market. Today, St Bartholomew's Hospital is at the centre of 900 years of healing and medical research and development.

LONDON LANDMARK

ST PAUL'S CATHEDRAL

St Paul's Churchyard, London, EC4M 8AD
www.stpauls.co.uk

There have been five churches on the site of St Paul's Cathedral, dating from AD 604. Sir Christopher Wren designed the latest version following the Great Fire of London in 1666. It was the first cathedral to be built in England and was completed within Wren's lifetime.

Christopher Wren was a mathematician and not a trained architect. He was born in East Knoyle in Wiltshire in 1632, the son of a clergyman. Initially, he went to Westminster School. Still, the Civil War was a disaster for the family as Wren's father lost his job twice. As a result, the family moved closer to Wren senior's brother near Oxford. As a young boy, Christopher Wren was influenced by his uncle, William Holder, a mathematician. He was tutored in geometry and later worked as a laboratory assistant to Charles Scarbrough, where he developed an interest in astronomy. Wren went to Wadham College, Oxford, in 1649, and in 1657 became the Professor of astronomy at Gresham College. He worked with other members of the Royal Society, including Robert Hooke, Robert Boyle, and Isaac Newton, to redesign London.

Tips for Visiting

The cathedral has a pre-booking arrangement, and there are also walk-in tickets at off-peak times. Pre-booking guarantees you an entry time if the building is busy. There is an entry charge but entering to attend evensong or pray is free. The building is closed to the public on occasions such as funerals or services of national importance, so checking the calendar on the website before visiting is advisable. Behind the ticket office are the volunteers who conduct tours of the cathedral floor during the day, included in the admission price (check the timings on the cathedral website). These are excellent insights into the history of the building, including the geometrical staircase, which is not normally open to the public. Tickets for tours of the Triforium are at an additional cost and should be pre-booked.

Scientific Highlights

The Geometric Staircase

Sir Christopher Wren constructed the geometric staircase as a testament to his mathematical genius. It is also known as the Dean's Staircase. It was designed to give the Dean access to the cathedral library. Steps lead from the Triforium to the cathedral floor, spiralling downwards. Remarkably, the staircase has also starred in films including one of the Harry Potter series and *Paddington 2*. Although it is hard to visualise, there are two flights of steps within the structure. The lower level is 6ft wide and ascends from the outer ground level to the cathedral floor and a landing level. The second flight of eighty-eight steps rises to the Triforium level without a landing level. Each step is 4ft wide and just a few inches thick. They appear to cantilever from the wall and are one of the world's longest continual flights of cantilevered stairs. Usually, a stair is fitted into place as the walls are constructed, which ensures the steps are securely built into the masonry. In 2005 repairs were carried out, revealing some of the original construction details. In 1704 the shaft was made. Each step was slotted

Dome of St Paul's Cathedral. (Author)

into it via a precise hole, creating the geometrical effect. Each lay directly on the one underneath and was held in place by metal wedges concealed in the mortar. The spiral staircase is one of St Paul's Cathedral's more dramatic and intriguing features.

The Dome

The dome at St Paul's Cathedral is a masterpiece but reliant on scientific precision and far more innovative than meets the eye. Christopher Wren was inspired by Rome and wanted to recreate something elaborate during the Restoration of the monarchy. The design is based on a triple dome structure that combines architecture with mathematics. From the exterior the magnificent hemispherical dome supports the lantern and dominates the skyline. However, Wren created two further domes. The middle layer is a concealed brick cone supporting the vast 850-ton lantern and outer dome. The inner dome rises above the circular cone of the cathedral and is visible from the inside. This structure enables the weight to transfer to the ground level of the cathedral. The outer layer is a wooden frame and covered with lead.

Naturally, Christopher Wren used mathematics to design the dome. He worked with his colleague Robert Hooke to calculate how to construct it successfully. Hooke drew on his theory of elasticity and tension (Hooke's Law). Hooke's weight experiments established that an unweighted chain suspended between two points would form a parabolic curve. Wren deduced that the converse was also true. A chain can hold unlimited weight in tension but will crumble if compressed. Brick, on the other hand, sustains compression but not tension. It is one of the reasons masons built flying buttresses into Gothic cathedrals to stop them from collapsing. Wren constructed the dome around a stone cone similar in shape to a parabolic arch. He inserted bands of steel chain at the base and around the dome's circumference to offset any problems with tension. The brick dome is just 9in thick but supports a lantern weighing 850 tons. Wren also knew the brick dome was not aesthetically pleasing, so it remained hidden. Instead, the inner dome was elaborately decorated by Sir James Thornhill and is admired by the public.

The Whispering Gallery

One of the most notable aspects of St Paul's Cathedral, the Whispering Gallery, was not originally intended as such but simply as a walkway around the dome. When Wren designed the area, the acoustics were not foremost in his mind. After the building was consecrated, the walkway became a fashionable meeting place. People began to hear more than they bargained for as the sound unexpectedly travelled around the gallery. A whisper causes a low-intensity sound wave, meaning there is less interference from other echoes or distortions. In 1870 Lord Rayleigh experimented in the Whispering Gallery. He deduced that sound travelled horizontally around the gallery by a reflective process. Sound waves can bounce around the circular dome multiple times because there are very slight angles. It explains why someone can hear a person whispering on one side of the gallery when they are on the opposite side. Whispering along the wall is more effective than whispering at it.

The American Chapel

The American Chapel was created in 1952 as a memorial to the United States servicemen who died in the Second World War. The stained glass window includes a symbol from each state, and the carvings depict native American birds and flowers. Try to spot the carved space rocket, formed as a testament to the significant scientific development in the 1950s in the USA, and poignantly, something the men who died never got to see.

The Crypt

Sir Christopher Wren's tomb is in the crypt with a simple epitaph that reads: 'Reader, if you seek his monument, look around you.' Memorials surround Wren's tomb, including his family, Robert Hooke, and masons and artisans who worked on the cathedral. The crypt also contains monuments to other scientists, including Sir Isaac Newton, Alexander Fleming, and Florence Nightingale.

SCIENTIFIC MEMORIALS AND STATUES

CHRIST'S HOSPITAL MEMORIAL, CHEAPSIDE

Newgate Street, EC1

A bronze memorial on the site of Greyfriar's Church commemorates 350 years of Christ's Hospital School's presence in the City of London. The school was founded by Edward VI in 1552 'to house, feed and educate needy children'. In 1673, Charles II founded the Royal Mathematical School, which was incorporated on the site. Isaac Newton supported the school. Robert Hooke designed a button for the boys to wear, which is still worn at Christ's Hospital School today. Boys from the school joined the Royal Navy using their mathematical knowledge to learn navigation skills.

Andrew F. Brown created the bronze sculpture following an open competition. The gardens within the ruined Greyfriars Church are a pleasant place to rest or enjoy lunch on the go.

Statue of Sir James Greathead in the City of London. (Author)

JAMES HENRY GREATHEAD STATUE

Cornhill, London EC3V 3NR

James Henry Greathead pioneered the construction of deep-level tunnels in the London Underground. A tunnelling implement, the Greathead Shield, is named after him. His statue is on an oval plinth in Cornhill but doubles as a ventilation shaft for Bank Station, one of London's deepest underground stations.

THE MONUMENT

The Monument, Fish St Hill, London, EC3R 8AH
www.themonument.org.uk

The Monument is a permanent reminder of the Great Fire of London in 1666 and is close to Pudding Lane, where the disaster first started. Designed by Robert Hooke with input from Christopher Wren, the Monument stands 61m high. Hooke had science in mind when he designed the structure and created a small laboratory beneath it. Both he and Wren intended to use a zenith telescope to observe the sky above the Monument. Unfortunately, the scientific experiments never materialised because the movement of the column made it unsuitable.

Visitors can climb the 311-step spiral staircase designed by Hooke and Wren to the top of the Monument for panoramic views across London. The steps are steep and narrow, so they are not for the faint-hearted. However, a certificate is awarded to those who complete the climb.

The City of London with St Paul's and the Monument. (Author)

LEARNING OPPORTUNITIES

GRESHAM COLLEGE LECTURES

www.gresham.ac.uk

Gresham College has been giving free public lectures since 1597, when Sir Thomas Gresham founded the institution to bring learning to the people of London. London's oldest higher education institution was the first university in England after Oxford and Cambridge. Christopher Wren was appointed Professor of Astronomy in 1657, and the alumni include Robert Hooke and Sir Isaac Newton. Today, the lectures remain free and are given on arts and science subjects by experts, including Professor Sarah Hart and Professor Sir Chris Whitty. Talks are held in venues across the City of London and are also live-streamed.

BARTS AND QUEEN MARY SCIENCE FESTIVAL

www.qmul.ac.uk/whri/patient-public-engagement/barts-and-queen-mary-science-festival

Each year the medical school at Barts and Queen Mary London host a science festival for older secondary school-age children and families. Topics range from genetics to cardiology and artificial intelligence and are free. Details are available from the William Harvey Research Centre website.

ARCHITECTURE WITH A SCIENTIFIC CONNECTION

MATHEMATICAL TILES

74 Long Lane, London, EC1A 9ET

The area around Long Lane and Cloth Fair in Smithfield has a lot of medieval architecture. No. 74 Long Lane has one of London's best examples of mathematical tiles. These were used in the seventeenth and eighteenth centuries, mainly in South-East England. The origin of the name mathematical is unknown. However, it is believed to relate to the geometric or exact layout of the tiles.

Builders laid mathematical tiles on the outside of timber-framed buildings as an alternative to brickwork. The tiles were overlaid onto timber and were a cheaper alternative to bricks in refacing medieval houses. The top of No. 74 is laid with mathematical tiles, while the remainder of the frontage is brickwork.

THE BUILDING THAT 'MELTED' CARS

20 Fenchurch Street, London, EC3M 8AF
https://skygarden.london

The Walkie-Talkie, otherwise known as 20 Fenchurch Street, is one of London's most imposing commercial buildings. Construction was completed in 2014. It got its name from the distinctive shape resembling a walkie-talkie, but it shot to fame for a different reason. The building attracted attention during construction with a solar glare problem. If the sun shone directly on the building for two hours in the day, a concave mirror effect occurred. No. 20 Fenchurch Street has a curved shape, which created a concave mirror. Concave mirrors collect light directed at them and refocus rays on a point. Typical uses include reflective telescopes, and the principle is used to light the Olympic flame. In the case of the Walkie-Talkie, the light was refocused on the street below, causing heat spots. The building hit the headlines when significant paint damage was done to a car in direct range of the ray, leading to concerns

City of London skyline with the Walkie Talkie building. (Author)

that it could destroy vehicles. In addition, a journalist claimed to have fried an egg on the pavement. As a result, protective screens were installed.

The building is not fully open to the public, but visitors can pre-book the rooftop restaurant and a Sky Garden visit with panoramic views of the City. Tickets are released each Monday for free Sky Garden access.

THE GHERKIN – INSPIRED BY NATURE

30 St Mary Axe, London

The vegetable-shaped skyscraper known as the Gherkin was designed by Norman Foster and completed in 2003. It stands 180m tall and is the second tallest building in London. It was inspired by the Venus flower basket sea sponge (*E. aspergillum*), which lives at great depths in the ocean. The lattice-shaped structure and hollow exoskeleton of the sponge enables it to disperse forces from strong currents, and that design has been adapted within the Gherkin architecture. The building is also one of the most sustainable in London due to its natural ventilation method.

The Gherkin's tapered and cylindrical shape also help it deflect air currents, so it has very little wind resistance. Older Brutalist-style square or rectangular buildings, such as Centrepoint in London, act differently by pushing air down to the pedestrian level, creating wind gusts. The Gherkin wind loads are minimised, making the building comfortable for people on the ground (bars and restaurants are at ground-floor level).

The differences in pressure created by deflecting the wind also contribute to the building's natural ventilation system. Space between the floors allows air to circulate naturally.

Inside the building, a lattice-like structure allows natural light and more open spacing. The building consumes half as much energy as other buildings of its size. The Gherkin is not usually open to the public, but it is possible to visit the Helix and Iris bars on the top floors, with panoramic views of the City.

THE BILLINGSGATE ROMAN HOUSE AND BATHS

101 Lower Thames Street, London, EC3R 6D
https://www.thecityofldn.com/directory/billingsgate-roman-house-and-baths/

The Billingsgate Roman House and Baths were first built around AD 150 and remained in use until the fifth century. Today, the ruins lie beneath an office block in the City of London, but they are a reminder of the remarkable engineering feats of the Roman Empire. The Romans used science to heat their baths. Many baths were symmetrical in design. Although early baths used thermal springs, the Romans later developed sophisticated heating systems such as under-floor (hypocaust) heating fuelled by wood-burning furnaces. The massive fires sent heat under the raised floors in the baths. By inserting hollow tubes into the masonry, the walls were also heated. Bosses in specially constructed bricks increased insulation and heat.

The Billingsgate Roman House and Baths can only be visited by guided tour from April to November (booked via the website).

MEDICINE AND HEALING

The fountain in the square at St Bartholomew's Hospital, London. (Author)

ST BARTHOLOMEW'S HOSPITAL AND MUSEUM

West Smithfield, London, EC1A 7BE
www.bartshealth.nhs.uk/bartsmuseum

There have been 900 years of healing at St Bartholomew's Hospital. The site was founded in 1123 by Rahere, a royal jester and courtier to Henry I. While on a pilgrimage to Rome, Rahere caught malaria. He had a dream where St Bartholomew appeared during his illness, asking him to establish a monastery and centre of healing in a place known as the Smooth Field. On recovering, Rahere returned to England and built a priory and hospital at Smithfield in London. During the twelfth century, the entire complex was a monastery and hospital combined, expanding to the area beyond Little Britain. Today, the hospital is one of the largest cardiovascular centres in Europe and has world-renowned expertise in cancer. It was here that William Harvey first conducted his experiments on blood circulation in the seventeenth century. Surgeon Percival Pott was the

The Hogarth Stair at St Bartholomew's Hospital. (Author)

first to identify environmental disease by recognising the connection between scrotal cancer and soot in chimney sweeps. James Paget was one of the founders of modern pathology and discovered Paget's Disease. Matron Mrs Ethel Gordon Fenwick pioneered nursing registration, which commenced in 1919. She is still remembered as State Registered Nurse No. 1.

The historic gateway has the only public statue of Henry VIII in London above it and dates from 1702. The main square at St Bartholomew's was designed by James Gibbs in the 1730s and is the focal point of the old hospital. The hospital added a magnificent fountain and gardens in 1859.

Within the Henry VIII gateway is the entrance to St Bartholomew's Hospital Museum. There are eclectic displays about the history of the hospital, old uniforms, and surgical instruments. The old hospital malt barrel is also on show, and a plaque commemorating the meeting of Sherlock Holmes and Dr Watson in the pathology laboratory.

The hospital's Grade I listed north wing contains the famous Hogarth Stair. William Hogarth lived in the area and painted 'The Pool of Bethesda' and 'The Good Samaritan' as a gift to the hospital. Some of the people in the paintings have distinctive medical symptoms, and even today, students are taught from Hogarth's masterpieces. The stair leads to the Great Hall, not normally open to the public, but is an ornate room initially built for the hospital management. It is decorated with lists of benefactors dating from 1546 to 1905, and some present-day donors, and the ceiling is gold leaf. Today it is used for ceremonial occasions.

ST BARTHOLOMEW THE LESS

West Smithfield, London EC1A 9DS

Until 2015, the small church of St Bartholomew the Less was unique as the only parish within a hospital. It is now part of the Parish of St Bartholomew the Great. The church dates from 1123 and has been on its

Stained glass memorial window to nurses from Barts Hospital who gave their lives in the Second World War. St Bartholomew the Less Church. (Author)

current site since 1184. However, the church became crown property during the Dissolution of the Monasteries in 1539. Henry VIII established the parish in 1547. The tower dates from the fifteenth century, with bells dating from 1380 and 1420. The stained glass windows were designed by Hugh Easton and dedicated in 1951. One is a memorial to nurses who died in the Second World War. Today, the church is used by staff and people using the hospital and is open to the public.

ST BARTHOLOMEW'S HOSPITAL PATHOLOGY MUSEUM

3rd Floor Robin Brook Centre, St Bartholomew's Hospital, West Smithfield, London, EC1A 7BE
www.qmul.ac.uk/pathologymuseum/about/index.html

The pathology museum at St Bartholomew's Hospital was the first medical museum established in London. In 1726 surgeon John Freke collected kidney stones from patients in exchange for medical fees, starting the collection of specimens studied by students and doctors eager to learn about medicine. The current Victorian museum was opened in 1879 and houses over 4,000 specimens in three mezzanine galleries with more storage. It is one of the largest collections in Britain. Notable examples include the skull of John Bellingham, who was executed for assassinating Prime Minister Spencer Perceval in 1812.

The museum is not open to the general public but opens for special events and by appointment only.

ST BARTHOLOMEW THE GREAT

The Priory Church of St Bartholomew the Great, West Smithfield, London, EC1A 9DS
www.greatstbarts.com

The church of St Bartholomew the Great is one of London's most atmospheric buildings and steeped in history. It is the oldest parish church in the City of London, dating from 1123, and was built as part of the priory by Rahere following his dream in Italy. The different architectural styles give just a little insight into Romanesque, Gothic, and Victorian features that make up the architecture. Rahere's tomb lies in the church and was once a site for pilgrims looking for healing. In 1539 the priory was closed due to the Dissolution of the Monasteries. It was restored to a parish church under the reign of Elizabeth I. The scientist and political leader Benjamin Franklin had a printing press in the Lady Chapel. Today, the church is used for worship and displays art such as Damian Hirst's sculpture Exquisite Pain. It is also reputed to be haunted by the ghost of Rahere.

Church of St Bartholomew the Great. (Author)

The church of St Bartholomew the Great is used frequently for filming, and credits include *Four Weddings and a Funeral* and *Shakespeare in Love*. Visitors can explore the church independently or pre-book a guided tour with the church staff (highly recommended).

MUSEUM OF THE ORDER OF ST JOHN

St John's Gate, St John's Lane, Clerkenwell, London, EC1M 4DA
https://museumstjohn.org.uk

This tiny museum is a delight to find and is one of London's hidden treasures. The original gatehouse was built in 1148 as the entry to the Grand Priory of the Order of St John of Jerusalem, and was burnt to the ground by Wat Tyler in 1381. It was restored to its present-day form by Prior John Redington before being rebuilt to its present format in 1504 by Prior Thomas Docwra. The Hospitallers, or Order of St John, were a religious order founded in Jerusalem around 1099, and dedicated to nursing and defence of the Holy Land.

The museum tells the story of the historic priory established in the 1190s and how the Order of St John and the Hospitallers evolved into the St John's Ambulance Brigade today. The museum contains paintings, tells how the Eye Hospital was established in Jerusalem and gives a history of the building. The museum is on the ground floor of St John's Gate. Visiting other parts of the building, including the crypt and priory church, is by guided tour only.

Museum of the Order of St John. (Author)

BRITISH RED CROSS MUSEUM

44 Moorfields, London, EC2Y 9AL
www.redcross.org.uk/about-us/our-history/museum-and-archives

This tiny museum in the British Red Cross Society headquarters has a fascinating collection of items from the organisation's history and current

British Red Cross Headquarters, London. (Author)

work. There are exhibits of medals and badges, medical equipment, and uniforms. One of the most compelling items is the Changi quilt, made by female Japanese prisoners of war.

BARBER SURGEON'S HALL

Monkwell Square, Wood St, Barbican, London, EC2Y 5BL
www.barber-surgeonshall.com

Barber Surgeon's Hall is the third building on this site and is an events centre. A Hall for The Worshipful Company of Barbers has been here since 1440. Barber surgeons were the precursors to modern-day surgeons and occupied themselves with bloodletting, boil lancing, and haircuts. Henry VIII decreed the Act of Union between the Company of Barbers and the Guild or Fellowship of Surgeons in 1540. A painting of the event by Hans Holbein hangs in Barber Surgeons Hall today, and there is a remarkable collection of lancets and bloodletting tools. The barbers and surgeons split in 1745 when the latter formed their own college, which became the Royal College of Surgeons. The Hall is not open to the general public apart from designated heritage days and events.

CUTLERS' HALL

4 Warwick Ln, London, EC4M 7BR
www.cutlerslondon.co.uk

The Cutlers' Company is one of the most ancient liveries in the City of London. It was first given a Royal Charter in 1416 and focused on swords and knives as the principal craft. However, as the need for surgical instruments increased in the nineteenth century, the company directed its attention to this trade. The building is not open to the public except for special events. The exterior of the building has a frieze showing the skills of the cutlers through the centuries.

Cutler's Hall, London. (Author)

APOTHECARIES HALL

Black Friars Ln, London, EC4V 6ER
www.apothecarieshall.com

The current Hall of the Apothecaries' Society is the oldest livery company Hall in the City of London still in existence. The apothecaries acquired their hall in 1632. In 1780 Apothecaries Hall underwent redevelopment with the addition of a laboratory for the large-scale manufacture of drugs. So when a young Agatha Christie sat her exams as a pharmaceutical assistant in 1917, she took them here. Today the building is used as an events centre and is only open to the public on defined heritage days.

PUBS WITH A PAST

THE GOLDEN BOY OF PYE CORNER AND THE FORTUNE OF WAR

Giltspur St, London EC1A 9DD

The gilt statue of a boy between Giltspur Street and Cock Lane marks where the Great Fire of London stopped in 1666. The child is overweight and designed to represent gluttony as a reminder the fire started in a baker's shop in Pudding Lane. However, this area hides a more sinister past. It is also the location of the notorious Fortune of War Inn, where resurrectionists or body snatchers used to gather to plot

Golden Boy of Pye Corner on Cock Lane, London. (Author)

thefts and sell corpses to the surgeons at St Bartholomew's Hospital across the road. At one time, the College of Surgeons rented rooms next to the inn where they practised dissection. In the eighteenth and early nineteenth centuries, surgeons needed to learn about anatomy and the human body. So they acquired corpses of executed criminals from Newgate Prison and unclaimed bodies from the workhouses for dissection. The resurrectionists, however, were supplying bodies from other sources. Corpses disappeared from graveyards. A local gang, known as 'burkers' after the notorious Edinburgh body-snatching duo Burke and Hare, not only stole from the local cemetery but preyed on the poor and needy, plying them with gin before killing them. The crime was finally detected when one of the surgeons received a young boy to dissect and noticed the body had no signs of burial. The boy was a young pauper from Smithfield Market. As a result, the gang was caught, tried, and condemned to death for their crimes. They were hanged at Newgate prison, close to today's Old Bailey. Ironically, their bodies were delivered to the surgeons for dissection. The crime caused a public uproar and led to parliament passing the Anatomy Act in 1832, which allowed doctors to dissect bodies that had been legally donated, finally regulating the procedure. The Fortune of War was demolished in 1919.

THE RISING SUN

38, Cloth Fair, London, EC1A 7QJ
https://risingsunbarbican.co.uk

The resurrectionists once used one of the most historic pubs in Smithfield, the Rising Sun, to lure unsuspecting victims for a drink, usually gin. Once under the influence of the concoction, they were despatched by the body snatchers. Bodies were also stored in the cellar before being sold to anatomists for dissection. As a result, it is reputed to be one of the most haunted pubs in London.

The Rising Sun, Cloth Fair, London. (Author)

THE HAND AND SHEARS

1 Middle Street, Cloth Fair, London, EC1A 7JA
https://thehandandshears.com

The Hand and Shears has been in existence since 1532 and is the only Grade I listed pub in Smithfield. Apart from having a reputation for serving excellent real ales, it played a part in the judiciary in the past. The Pye Powder Court was held here, which settled disputes between merchants and customers. Inquests, which examined the cause of death, were also presided over in the rooms above the bar.

The Hand and Shears pub. (Author)

THE OLD DOCTOR BUTLER'S HEAD

2 Mason's Avenue, Moorgate, London, EC2V 5BT
www.olddoctorbutlershead.co.uk

Dating from 1610, The Old Doctor Butler's Head is one of the most historic pubs in London. It was named after a seventeenth-century self-proclaimed specialist in neurological disorders who had some bizarre remedies. After a consultation for nervous disorders on London Bridge, Dr Butler suspended his patients through a trapdoor into the raging river below. To cure epilepsy, he fired pistols close to the patient to scare the disease away. Dr Butler was highly regarded and was appointed King James I's physician, despite not being qualified. He also developed a medicinal ale for gastric problems, sold in pubs he owned that displayed a Dr Butler's Head sign. Unfortunately, the current pub is the last one standing – and the medicinal ale is no longer served.

YE OLDE WATLING

29 Watling Street, London, EC4M 9BR
www.nicholsonspubs.co.uk/restaurants/london/yeoldewatlingwatlingstreetlondon

Located on one of Britain's most famous Roman roads, Ye Olde Watling was built by Sir Christopher Wren in 1688. The timbers are reputed to have come from ships. The building was constructed to house workers from the nearby St Paul's Cathedral project. The upstairs rooms were used as drawing offices.

THE OLD BELL TAVERN

95 Fleet Street, London, EC4Y 1DH
www.nicholsonspubs.co.uk/restaurants/london/theoldbelltavernfleetstreetlondon

This historic inn has changed names several times in its lifetime. Sir Christopher Wren originally built it in 1678 to house the stonemasons who were building St Bride's Church after the Great Fire of London.

THE ASTRONOMER

125-129 Middlesex Street, London, E1 7JF
www.theastronomerpub.co.uk

The Astronomer is a Victorian pub near Bishopsgate, inspired by the night sky. Astronomical clocks decorate the walls and there's space-themed music in the bathrooms. Among the cocktails and drinks in the Hubble Room, there's a Wray telescope and many astrological charts.

RESTAURANT WITH A SENSORY APPROACH

DANS LE NOIR

69-73 St John St, London, EC1M 4AN
https://london.danslenoir.com

Dans Le Noir is a popular restaurant with a difference. Dining is conducted totally in the dark, providing a unique sensory experience. Diners experience a gourmet menu in pitch darkness, enabling them to re-evaluate their taste and reclaim their senses. So what's on the menu? That's all part of the experience and mystery of the occasion.

The restaurant also offers sensory workshops, including wine tasting in darkness.

COFFEE BREAK

WREN CAFE

St Nicholas Cole Abbey, 114 Queen Victoria Street, London, EC4V 4B
www.thewrencoffee.com

The original church of St Nicholas Cole Abbey was destroyed in the Great Fire of London in 1666. Like many others in the City of London, it was rebuilt by the office of Sir Christopher Wren. Further damage in the Second World War led to additional restoration. Today, a popular coffee shop within the church sells drinks and light snacks, with all profits being used to maintain the building.

HISTORIC CHURCHES WITH A SCIENCE CONNECTION

ST OLAVE'S CHURCH

Hart Street, London, EC3R 7NB
https://saintolave.com

The entrance to St Olave's Church is formidable, with a series of skulls over the gateway. St Olave's is the burial site of some of the first victims of the Great Plague of 1665. During the Plague, it was said that churchyards rose in size by 3ft as a result of burying so many people, which terrified Samuel Pepys, who lived nearby. Today, there are steps down to the church building to reflect the added height. St Olave's is also where several medicinal plants are displayed. They include some of the 300 native plants that were

St Olave's Hart Street Church. (Author)

first described by the physician and naturalist William Turner (1508–68), famed as the father of English botany. Turner is buried in St Olave's Church with his son Peter, who was also a physician. Within the grounds is a small stone labyrinth.

The Seething Lane Gardens, adjacent to the church, are a quiet oasis in the City of London and contain references to events that happened in Samuel Pepys' lifetime. Of scientific interest are the paving stones depicting a scene from the Great Plague of 1665 and also a carving of a flea representing Robert Hooke's book *Micrographia*.

ST STEPHEN WALBROOK

39 Walbrook, London, EC4N 8BN
https://ststephenwalbrook.net

Engineers and scientists often work with prototypes before building their masterpieces. Sir Christopher Wren designed St Stephen Walbrook Church following the Great Fire of London in 1666, when most of the area had been destroyed. Wren designed the church, creating the dome and the ascending approach to the building. St Stephen Walbrook Church is believed to be the prototype for St Paul's Cathedral. The dome was the first to be built in England.

St Stephen Walbrook Church also has a plaque in memory of Nathaniel Hodge, the only doctor who remained with his patients in the parish during the Great Plague of 1665. Chad Varah started the Samaritans from this church in 1953, offering confidential phone help to people in severe mental distress who were suicidal. The original telephone is in the church.

ST MARY LE BOW

Cheapside, London, EC2V 6AU
www.stmarylebow.org.uk

St Mary Le Bow is most famous for the sound of its bells. They called Dick Whittington back to London in 1392 to become mayor. Being born within the sound of Bow Bells is the sign of a true Londoner or Cockney. However, this church also has scientific connections to Robert Boyle. Boyle (1627–1691) is most famous as one of the founders of modern chemistry and for his law on pressure and gas. He also wrote many philosophical works, including *Discourse of Things Above Reason* (1681). The terms of Boyle's will established a series of lectures to be given annually, bringing science and philosophy together. The first one was delivered by Richard Bentley in 1692. Each year a lecture is given on the connections with the natural world and Christianity by a scientist or theologian in the church.

St Mary Le Bow has self-guided tours of the church and a cafe in the crypt.

ST BRIDE'S CHURCH

Fleet Street, London, EC4Y 8AU
www.stbrides.com

St Bride's Church is famous for its spire and its association with journalists. The church was designed by Sir Christopher Wren after the earlier building was destroyed in the Great Fire of London in 1666. Church records show that the scientist Denis Papin was buried in the churchyard in 1713, and there is a memorial to him in St Bride's Church. Papin was born in France, forced to flee to Britain in 1675 due to his religious beliefs, and settled in London. He worked with Robert Boyle and is best known for his invention of the steam digester, which is the forerunner of the modern-day pressure cooker.

WEST END AND WESTMINSTER

Today the West End of London is famous for theatres and shopping, but there is a lot of scientific history in the area. For example, the Royal Institution is home to Michael Faraday's laboratory, where he conducted several of his historic experiments. Discover small museums with a wealth of heritage or find out how science applies to music and acoustics in the West End theatres. You'll also view Westminster with new eyes by learning where science legislation is made and the engineering behind one of the most famous clocks in the world.

LONDON LANDMARK

WESTMINSTER ABBEY

20 Dean's Yard, London, SW1P 3PA
www.westminster-abbey.org

Westminster Abbey is one of the most famous religious buildings in the world and is the burial place of many English and British monarchs. Since William the Conqueror in 1066, the abbey has hosted the coronations of kings and queens. Over 3,000 memorials are located within the abbey, which is famed for its Gothic architecture.

Tips for Visiting

Westminster Abbey gets busy, especially in peak times. Go early or late to avoid the crowds and pre-book tours such as the Queen's Diamond Jubilee Galleries area beforehand. Spend time exploring the cloisters in addition to the magnificent abbey, as there is a lot to see, from the Tomb of the Unknown Soldier to Poets' Corner, listening to a recital and the Quire.

Scientific Highlights of Westminster Abbey

Britain's oldest door

When visitors pass through the chapter house, the unremarkable door is probably the oldest of its kind in Britain. The oak has been dated

Westminster Abbey. (Coward Lion/Adobe)

using dendrochronology to the time of Edward the Confessor.

The Cosmati Pavement

The Cosmati pavement lies in front of the Great Altar in Westminster Abbey. Workers from Rome laid down the intricate design in 1268 on the orders of Henry III. The inlaid stone is called Cosmati work and uses cut work or an *opus sectile* technique. Although abstract with a four-fold symmetrical design, the flooring has geometrical patterns and other shapes, including triangles, rectangles, and squares.

Science Memorials

Many famous scientists are buried in or have memorials in Westminster Abbey. Several including Sir Humphrey Davy, Lord Rayleigh, and Thomas Telford are located in the North Ambulatory.

Near the altar in Westminster Abbey are the memorials to Charles Darwin and Isaac Newton. There are also memorials to Michael Faraday, James Clerk Maxwell, Ernest Rutherford, and Paul Dirac. The burial place of Stephen Hawking's ashes is in the nave, with memorials to David Livingstone and

Charles Lyell. John Hunter's memorial is in the north aisle of the nave.

The memorial to Lord Joseph Lister is in the North Quire Aisle.

The RAF Chapel

The RAF Chapel is located behind the tombs of Henry VII and Elizabeth of York. Just by the altar rail is the memorial to Sir Frank Whittle (1907–96). Whittle invented the jet engine, which revolutionised manned flight. The first test run took place in 1937, and by 1944 the RAF was using Meteor jet aircraft. Whittle died in 1996.

Poets' Corner

Many famous poets have memorials in this part of Westminster Abbey. However, there is also a memorial to Stephen Hales (1677–1761). Hales was a botanist who invented the ventilator, which helped people breathe. He was also the first person to measure blood pressure and determined that plants needed a component of air to breathe (carbon dioxide). The Halesia group of plants is named after him.

The Cloister

The cloister area is a beautiful place to reflect and contemplate the beauty of the abbey and surrounding area. There is also a memorial to the astronomer Edmond Halley in the south cloister. His achievements included mapping stars in the Southern Hemisphere and discovering Halley's Comet.

Memorial Window to Isambard Kingdom Brunel

On the south side of the abbey is a memorial window dedicated to Isambard Kingdom Brunel. It was initially installed on the north side in 1868 and then moved to its current position in 1952. The window depicts six biblical scenes and has the initials IKB at the top of the pictures. An inscription reads: 'In memory, Isambard Kingdom Brunel born April 9th 1806: departed this life September 15th 1859.'

The Nurses' Memorial Chapel

The Nurses' Memorial Chapel in the upper Islip chantry chapel is dedicated to Florence Nightingale. It was designed by Sebastian Comper in 1950 and is also dedicated to the nurses who died in the Second World War. The stained glass window was designed by Hugh Easton and there are badges from all the countries where the nurses came from. The lamp used in the annual memorial service to Nightingale is kept in the chapel. Visitors need to pre-book a visit to this chapel with the abbey vergers.

HOUSES OF PARLIAMENT, WESTMINSTER

Westminster, London, SW1A 0AA
www.parliament.uk/visiting

The Parliament of the United Kingdom is at Westminster and has several connections with science. Within the Palace of Westminster are two houses,

Westminster and the Elizabeth Tower. (Author)

the House of Lords and the House of Commons, that work on behalf of UK citizens to make and shape laws and challenge Government work. The iconic building on the River Thames dates from the eleventh century and has seen many changes and historical events. It has also been a royal residence, and the area outside the Palace of Westminster is where Charles I was executed. Following a fire in 1834, the Houses of Parliament were redesigned and restored under the leadership of Sir Charles Barry. Scientists, including Michael Faraday and Sir Goldsworthy Gurney, advised on the rebuilding.

Visitors can take a guided tour of the Houses of Parliament or a self-guided audio tour. These visits must be pre-booked, and visitors will experience significant security checks when entering the building. The tours cover the highlights of the building and explain how Parliament works.

Scientific Highlights of the Houses of Parliament

Westminster Hall

Most tours start in Westminster Hall, which is the oldest part of the Palace. The hammer-beam roof is a magnificent example of medieval engineering. A hammer beam is a type of timber truss that allows the roof to span longer than any length of timber. The hammer-beam roof in Westminster Hall is the largest in Europe. It was built between 1395 and 1399 and spanned 20.8m. The opening between the hammer beams is 7.7m.

New Dawn

Above the steps leading from Westminster Hall to St Stephen's Hall is a beautiful glass artwork called 'New Dawn' by Mary Branson. It was installed in 2016 and celebrates women's suffrage. Some women gained the right to vote in 1918, with all women being able to vote in 1928. The women's suffrage movement paved the way for equal rights for women, including access to science careers.

Central Lobby

The Central Lobby connects the House of Lords with the House of Commons and is symbolic because it is a meeting place for both houses. Laws including those affecting science, health and technology are made after hearings in the House of Commons and House of Lords before being given Royal Assent. The Central Lobby is neo-Gothic in design and was designed by Sir Charles Barry and Pugin.

The Peers Corridor connects the Central Lobby to the House of Lords, including the two arches to the left and right, which are used as voting corridors when laws and issues are being voted upon.

House of Lords

The House of Lords is the second chamber of Parliament and complements and is independent of the work of the House of Commons. Both Houses share responsibility for law-making. Much of the House of Lords work is conducted in committees that review public policy, including science and technology. Some of the House of Lords' members have expertise in health and science.

House of Commons

The Commons Chamber leading off the main House of Commons contains statues and busts of prime ministers from the twentieth century. Among them is Clement Atlee, whose Government founded the National Health Service in 1948. The House of Commons is the main chamber for debates, for Prime Minister's Questions, where issues can be raised, and for hearing new bills and laws. Commons Select Committees are also set up to review public policy and issues, including health and science.

ELIZABETH TOWER AND BIG BEN

Westminster, London, SW1A 0AA
www.parliament.uk/visiting

Probably the most famous timepiece in the world, Big Ben is the name given to the bell that tolls when the Great Clock of Westminster strikes. It lies in the Elizabeth Tower, originally called the Clock Tower but renamed to mark the diamond jubilee of Queen Elizabeth II in 2012. However, the entire tower and clock are commonly referred to as Big Ben. The Elizabeth Tower, the Great Clock, and Big Ben have undergone a major restoration programme. Today, visitors can book a guided tour of the Elizabeth Tower and Big Ben. The tour includes climbing the 334 steps, observing the clock mechanism, walking past the iconic glass clock faces, and standing next to Big Ben (with ear plugs) as it strikes.

Charles Barry was charged with building the Clock Tower, but he was not a horological expert. Before the nineteenth century, most timekeeping had been done using lunar methods. However, the construction of the tower at Westminster called for something unique. The Astronomer Royal George Airy was chosen to judge a competition for the most accurate turret clock in the world. Among many requirements for the timepiece, Airy specified the clock had to be accurate within one second each hour. The designs were created by Edmund Beckett Denison, who was an amateur horologist, and the mechanism was built by the watchmaker Edward John Dent. Tests to confirm the accuracy of Big Ben were conducted at the Royal Observatory in Greenwich, home to Britain's most accurate chronometers. Using a telegraph wire, time was sent from the timepieces at the Royal Observatory to Westminster to test the clock.

The clock maintains accuracy through a pendulum and cog system. The clock mechanism is a flatbed design so any part can be easily removed. There are three separate gear trains. One, known as the chiming train, controls the bells. The striking train controls Big Ben. The going train drives the clock hands and also triggers the striking train each hour and the chiming train every fifteen minutes. Each of the three trains is driven by one of three weights using the force of gravity.

The gravity escapement and pendulum work together to keep the clock ticking. Denison designed a double three-legged gravity escapement for the Great Clock. The pressure of the gravity arm on the pendulum rod keeps the pendulum swinging regularly.

The clock took two years to build and was opened in April 1859, with official timekeeping commencing a month later on 31 May 1859.

If the clock was to chime, bells were needed. A Great Bell was commissioned and cast by John Warner's foundry near

Stockton-on-Tees in 1856. Four quarter bells were cast at his Cripplegate foundry and took two weeks to cool. Denison wanted to get the loudest sound from the bell, so a large hammer was made. The bell was tested until October 1857, when a crack appeared because the hammer was too heavy. The bell was melted down, and George Mears from the Whitechapel Bell Foundry cast the new one on 10 April 1858, which is still in place today.

Unfortunately, and despite being lighter, this new bell also cracked when struck. So it remained silent for four years until a solution was found by the Astronomer Royal. George Airy advised the bell could be turned by 90 degrees and struck on a different side and with a lighter hammer.

Big Ben also makes a distinctive sound. Scientists used laser Doppler vibrometry to measure the clock's sound and found that it is thicker and heavier than other similar-sized bells; it has a higher pitch than would normally be expected for its diameter. The toll is not a single sound but a series of distinct frequencies and pulsations occurring simultaneously.

SCIENTIFIC MEMORIALS AND STATUES

ADA LOVELACE BLUE PLAQUE

12 St James Square, London, SW1

Ada Lovelace (1815–52) was the daughter of Lord Byron and a mathematician. She is best known for her work on Charles Babbage's mechanical computer, the Analytical Engine. Lovelace was the first person to realise the machine had other functionalities beyond calculation, including computing. She is regarded as the first computer programmer.

Ada Lovelace blue plaque, St James Square. (Author)

ADA LOVELACE STATUE

9 Millbank, London, SW1

In 2022, a sculpture of Ada Lovelace was unveiled at 9 Millbank. It was created by Etienne and Mary Milner and shows Ada in bronze looking out from the seventh floor, although hard to see from the street below. The bronze punch cards displayed behind her contain a puzzle created by scientists. Can you solve it?

MICHAEL FARADAY STATUE

2 Savoy Place, London, WC2

The bronze statue of Michael Faraday (1791–1867) stands outside the Institution of Engineering and Technology, formerly known as the Institute of Electrical Engineers. The original marble statue is in the Royal Institution.

JOSEPH BAZALGETTE MEMORIAL

Victoria Embankment, London

Joseph Bazalgette's memorial is located on what is perhaps his greatest legacy. During the 1850s, Bazalgette was Chief Engineer of the Metropolitan Board of Work and tasked with improving the state of London's sewers in response to the Great Stink of 1858. He created a network of sewers for London, modelling the design on a quadrupled population, so the system

Michael Faraday statue, Savoy Place. (Author)

Joseph Bazalgette Memorial, Embankment. (Author)

could operate effectively. Bazalgette built the Victoria Embankment and also part of the Albert Embankment so sewage could be carried away from the city. He was also responsible for the development of four Victorian pumping stations to move sewage out of London, including Abbey Mills and Crossness. Bazalgette's work saved thousands of lives by preventing cholera epidemics and other diseases. The Bazalgette Memorial was designed by George Blackall Simonds and is a few metres from Hungerford Bridge and Golden Jubilee Bridge, opposite Northumberland Avenue.

SAMUEL PLIMSOLL MEMORIAL

Victoria Embankment, London

Before Samuel Plimsoll invented the load line, ships frequently went to sea dangerously overloaded and sank. Plimsoll was a politician and social reformer who worked to improve the lives of mariners. The load line was developed to mark a ship's waterline and ensure that it had sufficient freeboard (the height from the waterline to the main deck) to maintain buoyancy. Initially, the Plimsoll Line was a simple sign with a circle and horizontal line. However, additional marks were added to reflect the effect of different water densities and climates on loads. The memorial to Samuel Plimsoll was erected by the National Union of Seamen in recognition of his work to save the lives of mariners worldwide.

Samuel Plimsoll Memorial, Embankment. (Author)

FLORENCE NIGHTINGALE STATUE

Waterloo Place, St James, London

Close to the Crimea Memorial in Waterloo Place is a statue of Florence Nightingale, commemorating her work in the Crimean War. It was designed by Arthur George Walker and unveiled in 1915.

MILLICENT FAWCETT STATUE

Parliament Square, London SW1P 3JX

The Millicent Fawcett Statue is the only monument dedicated to women in Parliament Square and was designed by Gillian Wearing. Fawcett was a prominent suffragist who fought for the right of women to vote. Take a closer look at the statue. Around the edges are photographs of other women who campaigned for equality. Among them are several women who promoted sciences and medicine.

Louisa Garrett Anderson qualified as a surgeon aged 24 and went on to form the Women's Hospital Corps during the First World War. She was the chief surgeon at the Endell Street Military Hospital. Lydia Ernestine Becker was an amateur scientist as well as a suffragette. She had interests in biology and astronomy.

Elsie Maud Inglis was a Scottish doctor, suffragette, and teacher. She founded the Scottish Women's Hospitals and was the first woman to be awarded the Serbian Order of the White Eagle. Jessie Chrystal Macmillan was a suffragist, barrister, and the first female science graduate from Edinburgh University.

Millicent Fawcett Statue, Westminster. (Author)

ENDELL STREET MILITARY HOSPITAL PLAQUE

65 Endell St, London, WC2H 9AJ

The Endell Street Military Hospital was opened during the First World War to treat wounded soldiers. It was led by Dr Flora Murray and Louisa Garrett Anderson and entirely staffed by women. The 573-bed hospital focused on psychological care in addition to physical treatments and conducted research into gas gangrene and other conditions. A small plaque in Covent Garden commemorates the women's achievements.

John Snow pump. (Author)

JOHN SNOW PUMP

Broadwick Street, London, W1F 9QJ

John Snow's name is best known for his famous intervention during an outbreak of cholera in 1894. Snow mapped incidences of cholera deaths in the Soho area and identified that people using a water pump on Broad Street were dying from the disease. He removed the handle from the water pump, which prevented people from using it, and reduced the incidence of cholera in that area. The water supply to the pump was found to be contaminated with sewage. Snow is considered as one of the founders of modern epidemiology. He also pioneered the use of obstetric anaesthesia, most notably in Queen Victoria during the birth of her eighth child. When Victoria had anaesthesia, it paved the way for other women, and this method of pain relief became popular. The pump on Broadwick Street is a memorial to John Snow, who died in 1858 from a stroke aged just 45.

MARYLEBONE BLUE PLAQUES

Both Wimpole Street and Harley Street have long been home to medical clinics. Several buildings in the area have blue plaques signifying former residents

Ethel Gordon Fenwick blue plaque, 20 Upper Wimpole Street. (Author)

with a science or medical background. At 20 Upper Wimpole Street, a plaque commemorates the life of Ethel Gordon Fenwick (1857–1947), who reformed the regulation of nursing. Sir Frederick Treves, the doctor associated with the Elephant Man, lived at 6 Wimpole Street, while the neurologist J.S. Risien Russell lived at No. 44.

The geologist Sir Charles Lyell lived at 73 Harley Street from 1854 to 1875. Elizabeth Garrell Anderson, the first woman to qualify as a doctor in England, has a blue plaque at 20 Upper Berkeley

Frederick Treves blue plaque, 6 Wimpole Street. (Author)

Street. The surgeon Sir Jonathan Hutchinson (1828–1913) is commemorated at 15 Cavendish Square, and Sir Ronald Ross, who discovered that malaria is transmitted by mosquitoes, lived at 18 Cavendish Square. Sir Patrick Manson, considered the father of tropical medicine, lived at 50 Welbeck Street. There is also a blue plaque marking the site where Florence Nightingale once lived at 10 South Street, Mayfair.

JOSEPH LISTER MEMORIAL

Portland Place, London

The Joseph Lister Memorial was designed by sculptor Thomas Brock in

Joseph Lister memorial. (Author)

memory of Joseph Lister, 1st Lord Lister. It was unveiled in 1924 and is close to his former home at 12 Park Crescent. The memorial's base is Aberdeen granite, while the sculpture is bronze. At the front of the statue, a woman points her right hand to Lister, and a small boy is holding out flowers.

Joseph Lister (1827–1912) is known as the father of antiseptics. He was a surgeon who pioneered the importance of sterile surgery, saving thousands of lives.

MUSEUMS WITH SCIENCE, MEDICAL AND ENGINEERING HERITAGE

THE FARADAY MUSEUM

Royal Institution, 21 Albemarle Street, London, W1S 4BS
www.rigb.org/visit/faraday-museum

It's quite incredible to think about the many experiments and discoveries that took place in the basement of 21 Albemarle Street. Fourteen Nobel Prize winners worked at the Royal Institution investigating science. Michael Faraday lived and worked at the Royal Institution from 1812 to the end of his life, making prolific contributions to scientific discovery. Today, the Faraday Museum is on the lower ground floor of the Royal Institution and has a display of the actual magnetic laboratory where Faraday worked along with other scientists. His work on electromagnetism is a highlight, but there is more. Faraday also worked on experiments defining the liquefaction of gas for cooling purposes, the development of the electric motor, and the electric generator.

Faraday's boss was Humphry Davy, and the Royal Institution has a display celebrating his prolific work on discovering nine chemical elements. Davy is best known for his work on the miners' safety lamp and was also

Royal Institution. (Author)

thought to be the inspiration for Mary Shelley's character Frankenstein. Alexander Fleming's work on lysosome and the value of 'nasal snot' is displayed, as is Sir James Dewar's innovations on thermos flasks. Exhibits in the museum include John Tyndall's research on why the sky is blue, and his theories on the Greenhouse Effect.

The Royal Institution closes periodically for all-day and private events, so checking the website before visiting is advised.

HUNTERIAN MUSEUM

The Royal College of Surgeons of England 38, 43 Lincoln's Inn Fields, London, WC2A 3PE
https://hunterianmuseum.org

The Hunterian Museum has one of the finest collections of medical specimens in Britain and is named after John Hunter, an eighteenth-century surgeon and anatomist. Hunter was a prolific collector of anatomical and pathological specimens, which formed the basis of

The long gallery of Hunter's collection at the Hunterian Museum. (Courtesy of Hunterian Museum copyright of Hufton and Crow)

the museum's collections. A series of eight galleries traces the development of medical and surgical understanding and practice through the centuries. Organised chronologically, the journey begins with examples of ancient anatomical and medical knowledge before introducing John Hunter. A series of smaller rooms feature aspects of his personal life and work, including his outdoor laboratory in Earl's Court, and the two-sided property he built near Leicester Square, which contained his extraordinary museum. Interactive displays tell the story of how William and John Hunter established their anatomy studies.

A long gallery houses over 2,000 specimens collected by Hunter, many of which have backstories about the lives of the people affected by disease. There are also exhibits about antiseptics, the development of surgical instruments, and surgical gloves (invented in 1889 by William Halstead). The museum includes information on how wars helped the development of surgical procedures such as the pioneering work of Archibald McIndoe performing surgery on injured pilots. Developments in MRI scanning and robotic surgery are also included. In Room 10, the Transforming Lives videos show how surgery has developed and how lives are being saved through advancement in techniques. The original heart of Jennifer Sutton is displayed in the room with a video of both her and her surgeon talking about her transplant experience.

BENJAMIN FRANKLIN HOUSE

36 Craven Street, London, WC2N 5NF
https://benjaminfranklinhouse.org

There is minimal furniture in the Benjamin Franklin House; however, the history and architecture of the building and its occupants are most interesting. The house on Craven Street dates from the 1730s and is the world's only remaining home of Benjamin Franklin. The scientist and founder of the United States of America lived in the house

Benjamin Franklin House. (Author)

from 1757 to 1775. During his stay, Franklin conducted his famous kite experiment and invented the glass armonica.

The basement of 36 Craven Street was used as an anatomy school in the nineteenth century, resulting in several bones being excavated during the conservation of the house in 1998. One of the best ways of getting an insight into the house's history and Franklin's life is to pre-book a historical and architectural tour.

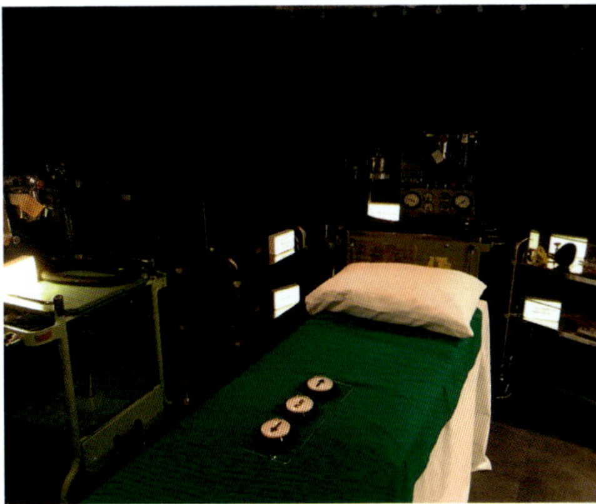

Interactive operating theatre at the Anaesthesia Heritage Centre. (Courtesy of the Anaesthesia Heritage Centre)

BRITISH OPTICAL ASSOCIATION MUSEUM

42 Craven Street, London, WC2N 5NG
www.college-optometrists.org/the-british-optical-association-museum

The small museum at the College of Optometrists is home to over 28,000 items relating to optics. Displays include historic visual aids, diagnostic equipment, and models of ocular disease. The museum prides itself on being the only place where the exhibits look at you.

ANAESTHESIA HERITAGE CENTRE

21 Portland Place, London, W1B 1PY
https://anaesthetists.org/Home/Heritage-centre

The Anaesthesia Heritage Centre is a small museum at the Association of Anaesthetists. It is packed with displays about the history of anaesthesia from 1846 through to the modern day, including many significant developments in medical history. The collections include ventilators and airway instruments, references to resuscitation, pain relief, and the historical aspects of anaesthetics. There is also an interactive display of equipment and machinery used in anaesthetics.

ROYAL COLLEGE OF NURSING LIBRARY AND HERITAGE CENTRE

20 Cavendish Square, London, W1G 0RN
www.rcn.org.uk/library

Europe's most extensive specialist nursing library is located at the Royal College of Nursing headquarters.

There is also a rich collection of historical archives relating to nursing. Exhibitions are held throughout the year.

LONDON TRANSPORT MUSEUM

Covent Garden Piazza. London, WC2E 7BB
www.ltmuseum.co.uk

The London Transport Museum has a prolific collection of items and vehicles connected to the development of transportation in the capital. Exhibits include a display of a disused underground station (visitors can also book a Hidden London tour of decommissioned stations via the museum and see a site for real). A display outlining how the world's first underground was created in 1860 highlights the engineering that went into the development. Learn how the Greathead shield enabled deeper digging for new stations like Moorgate. The museum also has an interactive exhibition for engineers of the future. Aspiring engineers can solve traffic conundrums, use STEM (science, technology, engineering and mathematics) skills, use ticket technology, and learn how to plan greener transport.

The London Transport Museum has a depot at Acton where engines and other vehicles are displayed on open days and for guided tours.

SCIENCE ON THE STREET

GAS LAMP ON CARTING LANE

Carting Lane, London, WC2R 0DW

Just off the Strand, the delightful lantern on Carting Lane is an ingenious example of Victorian engineering. Engineers needed to find a sewer ventilation system and came up with the idea of sewer gases lighting street lamps. The sewer gas destructor lamp was patented by Joseph Edmund Webb in the 1890s. The lamp on Carting Lane is the last of its kind in the City of Westminster.

Gas lamp on Carting Lane. (Author)

EPSTEIN STATUES, STRAND

429, Strand, London

In 1907, the eminent sculptor Jacob Epstein received a prestigious commission from Charles Holden, the architect who designed the old BMA building on the Strand. Epstein was asked to create a series of sculptures. His creations were a series of male and female nude sculptures displaying the creativity and decay of life. However, although crowds flocked to see them, they caused public outrage, particularly from the church. In the 1930s, the Southern Rhodesia Government purchased the building, and the statues were mutilated in 1937 as they were thought inappropriate. Today, the building is known as Zimbabwe House.

Epstein statues on the Strand. (Author)

SCIENTIFIC FACTS IN TRAFALGAR SQUARE

Trafalgar Square is most famous for Nelson's Column and the lions, but several other scientific items are within plain sight.

STANDARD IMPERIAL MEASUREMENTS

In the north-east corner of Trafalgar Square are a series of metal plaques. Some are on the wall, and others are on the steps leading to the National Gallery. They depict imperial measurements set at 62° F and were established by the Board of Trade to protect standard measures. The plaques include inches, feet, perches, and others, and have been installed in Trafalgar Square since 1876.

GOLDSWORTHY GURNEY'S BUDE LIGHT

Trafalgar Sq, London WC2N 5DS

At the south-east corner of Trafalgar Square, the compact structure with a glass lamp is best known as Britain's smallest police station. However, the glass lantern was patented by Sir Goldsworthy Gurney in 1839 and named a Bude

Goldsworthy Gurney's Bude Light in Trafalgar Square. (Author)

Light after his home town in Cornwall. The lantern is a gas-fired lamp using pure oxygen and exuding a very bright light. Gurney showed the invention to Michael Faraday as a potential use in lighthouses. Although it was trialled, the lamps were not used for lighthouses due to the cost. However, they were modified with air and used to light the Houses of Parliament. Four Bude Lights were installed in Trafalgar Square in 1845. Today, the Bude Light works using electricity.

UNDERGROUND LONDON

DOWN STREET UNDERGROUND STATION TOUR

www.ltmuseum.co.uk

London Transport Museum operate several tours into disused underground stations with a significant history. Down Street is near Piccadilly and had a short life as an underground station between 1907 to 1932. However, its role in acting as the secret headquarters of the Railway Executive Committee during the Second World War is one of the more intriguing aspects of the tour. Apart from the engineering features of the tunnels, there are the remaining rooms used by railway staff to plan vital logistics across the country in wartime. Winston Churchill is also believed to have stayed in the underground station. From a science and engineering perspective, the ventilation shafts give insight into early designs and how tunnels were ventilated. In addition, there's a 1939 telephone exchange and insights into how sewage was extracted from the station. Visitors can also learn about and experience the physics of the piston effect of being in an underground railway station. When a tube train travels through a narrow tunnel, the positive pressure at the head pushes foul air out. The negative pressure at the rear sucks fresh air, generating a piston wind. The piston wind ventilates the station.

The Down Street Tour and other underground experiences can be booked via the London Transport Museum. The Moorgate underground experience has a complete Greathead shield dating from 1904 in addition to other highlights in one of London's deepest stations. Pre-booking is essential.

LEARNING OPPORTUNITIES

THE ROYAL INSTITUTION

Royal Institution, 21 Albemarle Street, London, W1S 4BS
www.rigb.org

The Royal Institution has a programme of workshops and talks for schools throughout the year, listed on their website. The RI Christmas Lectures are among the most famous events at the Royal Institution and were initially introduced by Michael Faraday when education for children was sparse. These lectures have continued since 1825 and feature eminent scientists giving talks to an audience aged 11–17. They have also been televised since 1966. Tickets are allocated by ballot with details of how to enter on the RI website.

ROYAL SOCIETY

6-9 Carlton House Terrace, London, SW1Y 5AG
https://royalsociety.org

The Royal Society is a fellowship of eminent scientists and is the

Michael Faraday lecturing at the Royal Institution: Prince Albert and his sons in the audience. Wood engraving, 1856, after A. Blaikley. (Wellcome Collection ref 40199)

oldest scientific academy that is in continuous existence. The first meeting occurred in 1660, and members have included Sir Robert Boyle, Sir Isaac Newton, Sir Humphrey Davy, Michael Faraday, and Dorothy Hodgkin. Today, the Royal Society has an extensive library that can be visited by pre-appointment and also hosts lectures and events on science.

THE LINNEAN SOCIETY

Burlington House, Piccadilly, London, W1J 0BF
www.linnean.org

The Linnean Society is the world's oldest society dedicated to the natural world. It takes its name from the Swedish naturalist Carl Linnaeus. The organisation has a range of events and workshops that can be booked via its website.

WALKING TOURS

COVENT GARDEN WITH A DIFFERENCE

https://unseentours.org.uk

Unseen Tours is a social enterprise running walking tours in London with a difference. Each tour is designed by and led by someone who has experienced homelessness in the city. You'll get an insight into some of the attractions in the area but will also hear first-hand about what it is like to be homeless. For example, Viv's walking tour includes Michael Faraday and the story of the women who rebuilt Waterloo Bridge in the Second World War.

THEATRICAL SCIENCE IN THE WEST END

LYRIC THEATRE

29, Shaftesbury Avenue, London, W1D 7ES
https://thelyrictheatre.co.uk

The Lyric is one of the top West End theatre venues, but it has an interesting medical history. During the 1760s, William Hunter moved to the site on Great Windmill Street at the rear of today's theatre and established the Hunterian Anatomy School. A blue plaque commemorates the surgeon.

ROYAL OPERA HOUSE

Bow Street, Covent Garden, London, WC2E 9DD
www.roh.org.uk

There has been an opera house on the Bow Street site since 1732, but the building underwent a massive restoration in the 1990s when the theatre was modernised. Today, visitors can book a Behind the Scenes Tour and hear how the furnishings are designed to maximise acoustics and how lighting and special effects are managed. Other scientific insights include the ventilation system through the theatre and how costumes are cleaned and maintained. Naturally, the best way to appreciate all the science and special effects is to book a performance at this famous theatre.

BRIDGE OF ASPIRATION

Floral St, Covent Garden, London

The Bridge of Aspiration provides dancers with a direct link from the Royal Ballet School to the Royal Opera House in Covent Garden. It was designed by Wilkinson Eyre Architects and has distinctive geometrical features. The bridge has a concertina effect with twenty-three square portals supported by an aluminium beam. They rotate in sequence, so the bridge appears to have undergone a quarter turn. The design is aimed at portraying the fluidity and grace of dance.

Bridge of Aspiration at the Royal Ballet School, London. (Author)

ARTS AND SCIENCE

NATIONAL PORTRAIT GALLERY

St Martin's Place, London, WC2H 0HE
www.npg.org.uk

The National Portrait Gallery is the most extensive collection of portraits in the world, with over 215,000 pieces of art.

Science is integral to the work of the National Gallery in analysing paint and the structure of a painting. X-rays, gas and mass spectrometry, and infrared reflectography are some of the techniques used in analysis and conservation.

The gallery is home to several paintings featuring people associated with science and medicine. Some of the people with portraits in the gallery include Sir Isaac Newton, Ada Lovelace, Mary Seacole, Stephen Hawking, Maggie

Aderin-Pocock, Sir Joseph Banks, Sir Frederick Grant-Banting, Marie Curie, Charles Darwin, Sir Humphrey Davy, Sir Alexander Fleming, Michael Faraday, Joanna Dorothy Haigh, Marie Stopes, and Sir Christopher Wren.

STATUES AT THE ROYAL ACADEMY OF ARTS

Burlington House, Piccadilly, London, W1J 0BD

Burlington House is home to the Royal Academy of Arts and dates from the seventeenth century, with refacing being completed in the eighteenth century. Although primarily concerned with the arts, the building's exterior has several scientists on the lower and upper galleries. The sculptor John Durham created the statues above the doorway of William Harvey, Jeremy Bentham, and Isaac Newton.

Statue of William Harvey at the Royal Academy. (Author)

On the central balustrade are W.F. Woodington's sculptures of Galen, Archimedes, and Aristotle. Galileo is on the eastern balustrade. On the opposite side are Humphrey Davy and John Hunter. Leibnitz, Cuvier and Linnaeus by Patrick MacDowell decorate the ground-floor east side of the building.

HOTELS WITH A CONNECTION TO SCIENCE

BROWN'S HOTEL

Albemarle Street, London
www.roccofortehotels.com/hotels-and-resorts/brown-s-hotel

One of London's most famous hotels, Brown's, has hosted several prominent scientists and seen historical events. Brown's Hotel incorporated several neighbouring hotels in the 1880s, including St George's Hotel in 1889, which was known for hosting the famous X Club. The private dining club met between the 1860s and 1880s and consisted of nine prominent scientists, including Joseph Hooker and Thomas Huxley. All were believers in Charles Darwin's Theory of Natural Selection. In

1877, Britain's first telephone call was made by Alexander Graham Bell from Brown's Hotel. In 1890, Lord Kelvin led the International Niagara Commission meeting in the Niagara Room, where it was agreed the best way to distribute power from the hydroelectric plant at Niagara was agreed.

LANGHAM HOTEL

1c Portland Place, Regent Street, London, W1B 1JA
www.langhamhotels.com/en/the-langham/london

The Langham Hotel dates from 1865 and was opened as one of Europe's grand hotels. It was the first in the world to install a hydraulic lift system. In 1868, Thomas Evans stayed at the Langham when demonstrating the use of nitrous oxide as an anaesthetic. The hotel's award-winning cocktail bar, Artesian, is named after the 350-year-old artesian well under the building. Learn the art and science of cocktail making at one of the popular masterclasses run by Sauce, the Langham Cookery School. You'll master three classic cocktail techniques, learn why ice quality is essential, and see why some cocktails need to be shaken and not stirred. You'll also get a behind-the-scenes look at how cocktails are created in the Artesian's lab.

Cocktail making at the Langham Hotel. (Author)

RESTAURANTS, PUBS AND BARS CONNECTED TO SCIENCE

THE JOHN SNOW

Broadwick Street, London, W1F 9QJ

A quaint pub adjacent to the John Snow pump on Broadwick Street and named after the famous epidemiologist.

THE ALCHEMIST

63-66 St Martins Lane, London, WC2N 4JS
https://thealchemist.uk.com

A fun cocktail bar with drinks themed on molecular magic all add to the alchemic atmosphere in this West End bar. There's a restaurant, too, with internationally inspired cuisine.

MR FOGG'S APOTHECARY

Brook Street, Mayfair, London, W1K 5DN
www.mr-foggs.com/mr-foggs-apothecary

Designed to look like a Victorian townhouse and pharmacy, Mr Fogg's Apothecary is a cocktail bar with a range of elixirs and tinctures themed on a medicinal and healing basis.

3

BLOOMSBURY

Bloomsbury lies at the centres of intellectual and cultural London. It is home to the University of London and some of London's famous teaching hospitals, forming a natural environment for scientific development. The British Museum has one of the most extensive collections in the world, including many scientific items. There is engineering, too, with innovative construction on Regent's Canal and the railways bringing people and goods to the capital.

LONDON LANDMARK

THE BRITISH MUSEUM

Great Russell Street, London, WC1B 3DG
www.britishmuseum.org

One day does not do the British Museum justice. As one of the world's most famous museums, it has a vast collection of exhibits, and several have scientific connections. The museum came about as a result of an Act of Parliament in 1753 to create the world's first national, public and free museum.

The British Library. (© visitlondon.com/ Jon Reid)

It was opened in 1759. Sir Hans Sloane, an eighteenth-century physician and member of the Royal Society, amassed a collection of over 80,000 'natural and artificial rarities' along with a vast library of books, which were included in the museum's acquisition.

Tips for Visiting

Entry to the British Museum is free, but some special exhibitions have a charge. The entrance area in the Great Court has help desks that can direct you to specific rooms, and there is a helpful map with suggestions for an hour or half a day of visiting. The bookshop is also worth a visit.

Scientific Highlights of the British Museum

The Geometry of the Great Court

The Great Court is the starting point for most visitors to the British Museum. In 2000, Norman Foster's design of the Great Court and Reading Room was opened to the public and allowed people to move freely around the ground floor. The Great Court is the largest open enclosed space in Europe. However, the roof has geometric features within its architectural design. It is made of 3,312 glazed triangles, none of which are the same. In addition, the drum of the reading room is off-centre, so the roof canopy had to reconcile the asymmetry with the rectangular courtyard. Computer-generated technology created a gently curved glass roof, which enabled the steel structures to meet within 3mm accuracy. The result is an aesthetically pleasing space and an iconic feature in the central part of the museum.

Room 1 Hans Sloane Collection

Sir Hans Sloane was a prolific collector. Many items were donated to the British Museum and are displayed in Room 1. Exhibits include a range of fascinating scientific instruments, including an orrery to measure the solar system and the position of the planets. The collection also contains telescopes, a pair of prisms, a model sucking pump, the Sloane astrolabe, and navigational equipment. There is also a rock and mineral collection, including fossils.

Rosetta Stone

The Rosetta Stone is one of the most famous items in the British Museum. It was discovered in 1799 by Pierre-Francois Bouchard during Napoleon's Egyptian Expedition. The Rosetta Stone held the key to understanding Egyptian hieroglyphics.

In 1801, the British Army defeated the French in Egypt, and the Rosetta Stone was passed to the British Museum, where it has been on display since 1802. Jean-François Champollion, a Frenchman and son of a bookseller, developed a passion for Egypt. In 1802 he met the mathematician Joseph Fourier, who had recently returned from Egypt and had several items in his possession with hieroglyphics. Champollion was so inspired he decided to decipher the hieroglyphics. Twenty years later, following work on similar hieroglyphic tablets and journeying in Egypt, Champollion had managed to crack the code behind the Rosetta Stone, which led to a transformation in the understanding of Egyptian history and hieroglyphics.

Islamic Astrolabes

Astrolabes were used to solve complex problems, such as being able to identify the position of the stars, constellations, and other heavenly bodies. These instruments have several moving parts and use raw data. They were the computers of their age and were used for timekeeping, determining the position of Mecca, and the time for prayer. A qibla is also within the collection and was used to determine the position of Mecca from twenty cities. Examples of astrolabes are displayed in the Albukhary Foundation Galleries of the Islamic World.

Lindow Man

Lindow Man was found in a peat bog in Cheshire in 1984. He is one of the most prominent exhibits in the British Museum. The peat bog preserved his body well. Peat bogs contain sphagnum moss, and a sugary substance called sphagnan is released when they die. It is a natural tanning agent, turning skin into leather and giving hair and bodies a brown colour. Lindow Man has undergone many scientific tests to determine his age, characteristics, and how he died. Examples include xeroradiography and CT scanning to find out more about his appearance. Scientific methods such as freeze drying were also utilised to preserve Lindow Man and prevent decay.

Automaton

The mechanical galleon or automaton dates from 1585 and was built by Hans Schlottheim. The automaton is also a clock and works on three intricate spring mechanisms. It has many detailed features and was used to summon people to dinner rather grandly. It is in Room 39.

Mechanical Clocks

There is a large display of mechanical clocks in the British Museum, which includes how a clock needs energy to work. Exhibits include clocks from the sixteenth century and highlight the accuracy and precision required by clockmakers.

Cradle to Grave Pharmacopoeia

This captivating art installation by artists Susie Freeman and David Critchley and family doctor Liz Lee show the type of medication taken by the average man and woman in their lifetime. Two knitted panels or pill diaries set out the types of medication taken in chronological order, culminating in more than 14,000 pills. The installation is 14m long and is located in the Wellcome Trust Gallery.

LONDON LANDMARK

LONDON ZOO

Regent's Park, London, NW1 4RY
www.zsl.org/zsl-london-zoo

London Zoo was opened in 1828 and is the world's oldest scientific zoo. It was initially intended to be a place to study science. In 1832, the animals from the menagerie at the Tower of London were transferred to London Zoo, which was opened to the public in 1847. Today, the zoo has over 14,900 animals on the Regent's Park site. London Zoo is run by the Zoological Society of London (ZSL), a registered charity focusing on science and animals, including conservation.

Tips for Visiting

Visitors must pre-book a general admission ticket and any experiences. However, the zoo operates a membership scheme, and members do not need to pre-book general admission tickets.

There is a lot to see at London Zoo, and more than one visit is needed

Giraffes in London Zoo, Regent's Park. (kmiragaya at Adobe)

to see everything. Instead, focus on a few highlights and check the zoo website for any special events, such as a talk. If you have a favourite animal, include them in your day at the zoo.

Experiences

London Zoo operates several exciting experiences that must be booked in advance. They include behind-the-scenes tours with a keeper to see Komodo dragons or giraffes, penguin feeding, and meeting meerkats or monkeys. There are also sensory experiences for people with special needs and a spider programme for people with arachnid phobias.

London Zoo has lodges where guests can stay overnight and wake up to the sounds of lions and other animals.

Highlights

There are some wonderful ways to see the animals at London Zoo. The zoo is divided into pink, orange, and blue zones. Rainforest World is an indoor display area within the pink zone and includes a nocturnal experience. You'll also find giraffes, African wild dogs, and meerkats. The Monkey Valley is another popular area. Within the orange zone are lions, llamas, penguins, and camels. Finally, the blue zone is home to the iconic reptile house, tigers, Komodo dragons, and gorillas.

LONDON LANDMARK

ST PANCRAS INTERNATIONAL STATION

Euston Rd., London N1C 4QP
https://stpancras.com

St Pancras was built in 1868 by the Midland Railway Company, providing a vital connection between London and some of England's prominent cities. It is also a fine example of Victorian engineering and Gothic architecture.

Engineers William Henry Barlow & Rowland Mason Ordish were appointed to design the overall station layout, train shed and track design. George Gilbert Scott was responsible for creating the station accommodation and the hotel architecture, which began in 1876. Today the station connects Britain to Europe. It has been significantly upgraded and is one of the most elegant stations in the world.

One of the significant challenges for Barlow and Ordish was how to cross the Regent's Canal with a rail track. So instead, they constructed the train shed platform deck 5m above the ground, which provided space for storage

Interior of St Pancras International – a central London railway terminus on Euston Road in the London Borough of Camden. (Adrian Popescu at Adobe)

below, unlike many other stations at the time. That space was used by local brewing companies to store beer.

When the Barlow-designed iron and glass roof was installed at St Pancras Station, it was the largest single-span roof in the world, at 75m, and it was 240m long and 30m high. Twenty-five principal arched trusses constitute the roof arches and are linked by longitudinal purlins, which rise to a slight point at the crown. The design has been used worldwide in many stations, including Grand Central Station, New York. Today, the roof also has an environmental function as it is home to several beehives.

Railway Time

Timekeeping is synonymous with railways. Before the development of the railways, timekeeping was based on the rotation of the earth, which resulted in small but significant time differences between London and Bristol, and confusion with train timetables. In the 1840s, Railway Time evolved, which was based on London Time and the timings at the Royal Observatory in Greenwich. Clocks in railway stations became important, and the original timepiece installed on the southern end of the train shed in St Pancras Station was reputed to be the largest in any railway station in England. The diameter was 5.15m, and this magnificent clock remained in place until the 1960s when British Rail decided it needed modernising. As it was lowered, the clock was dropped and broke into several pieces. One of the train guards, Roland Hoggard, offered to buy the clock for £25 and took it to his home in Nottinghamshire, planning to restore it. British Rail took casts of the clock, and made a plastic replica with upgrades to the mechanism. This clock remained at St Pancras until the station transformation in the 2000s when Hoggard's clock story was recalled. Templates were taken of the original clock, and a new one was created by Dent with Smith of Derby, two companies with a strong clockmaking history. The third clock keeps excellent time and is in pride of place in the train shed.

SCIENTIFIC MEMORIALS AND STATUES

LOUISA BRANDRETH ALDRICH-BLAKE

Tavistock Square Gardens, Tavistock Place, London, WC1H 9RE

Louisa Brandreth Aldrich-Blake (1865–1925) was one of the first women to study medicine and was the first female surgeon in Great Britain. She held several prominent positions, including the first female anaesthetist at the Royal Free Hospital in 1895. During the First World War, she worked in field hospitals in France and was known for her meticulous work on ensuring all traces of ordnance were removed

Statue of Louisa Brandreth Aldrich Blake in Tavistock Square. (Author)

from wounds. From 1914 she was appointed Dean of the London School and increased the number of women accessing medical careers.

HODGKIN AND WAKELEY, BEDFORD SQUARE

35 Bedford Square, London, WC1B 3ES

Blue plaques in Bedford Square mark the one-time residence of two medical pioneers. Thomas Hodgkin (1798–1866) was a physician and philanthropist who discovered Hodgkin's Disease. Thomas Wakeley (1795–1862) founded the medical journal *The Lancet.*

Thomas Wakeley and Thomas Hodgkin blue plaques. (Author)

INSECTS AND PARASITES AT THE LONDON SCHOOL AND TROPICAL MEDICINE

Keppel Street, London, WC1E 7HT
www.lshtm.ac.uk

Look up when you walk past the London School of Hygiene and Tropical Medicine. The doors and balconies are decorated with reliefs of various insects and parasites that cause diseases.

SCIENCE AND MEDICINE IN GOWER STREET

Gower Street runs from Euston Road to Montague Street and has several interesting plaques relating to science on its buildings. The first anaesthetic in England was administered at 24 Gower Street in 1846. James Robinson (1813–62), a pioneer of dentistry and anaesthesia, lived at 14 Gower Street

Plaque on 12 Gower Street commemorating the first anaesthetic administration. (Author)

and administered ether to a person having a tooth extraction. At 12 Upper Gower Street (later 110 Gower Street), Charles Darwin rented the property in 1835, famously writing to his fiancée Emma Wedgwood and calling it Macaw Cottage because of the gaudiness of the décor. Sir Victor Horsley, known as the first neurosurgeon, lived at 129 Gower Street. In 1887 he was the first person to remove a tumour from a spinal cord. He was also instrumental in eradicating rabies from the United Kingdom. On the University of London wall is a memorial to Richard Trevithick (1771–1833), a pioneer of high-pressure steam. In 1808 he ran the first passenger-carrying steam locomotive.

SCIENCE IN THE LIBRARY

THE BRITISH LIBRARY

96 Euston Road, London, NW1 2DB
www.bl.uk

The British Library is an extensive repository for books, documents and manuscripts and has a reading room for researchers. Visitors can see some of the most famous scientific papers in the world in the Treasures of the British Library exhibition. Displays include Michelangelo's anatomical illustrations, one of the first photographs of the moon, Florence Nightingale's original Diagram of the Causes of Mortality in the Army in the East, and Leonardo da Vinci's notebooks detailing his inventions.

In the piazza at the British Library is a sculpture of Isaac Newton by Eduardo Paolozzi, thought to be searching for knowledge. Newton's body is made with nuts and bolts to demonstrate the connection between science and nature.

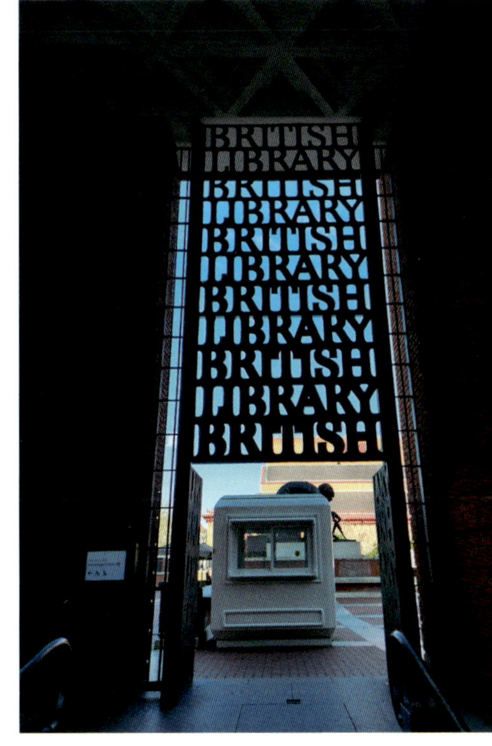

Entrance to British Library. (Author)

Within the amphitheatre is Anthony Gormley's Planets artwork. Each of the eight igneous rocks originated from Scandinavia between 350 and 1,000 million years ago. A small carved figure clings to each stone, demonstrating the connection between the environment and humankind.

Please don't leave the British Library before visiting the lower-floor cloakroom, where the Paradoxymoron by Patrick Hughes may confuse you with its optical illusion. It is a sculpted painting of bookcases using geometry with 90 and 45 degrees angles to make people think the structure is following their movements.

MUSEUMS WITH SCIENCE, MEDICAL AND ENGINEERING HERITAGE

THE LONDON CANAL MUSEUM

12-13, New Wharf Road, London, N1 9RT
www.canalmuseum.uk

The London Canal Museum is housed in a building originally used as a warehouse for ice imports for the ice cream trade in the nineteenth century. Exhibits show how the canals were built and transported goods during Victorian times to the Regent's Canal Basin outside the museum. The mitre gate, designed by Leonardo da Vinci,

Regent's Canal at the London Canal Museum. (Author)

is a feature of modern locks, and the museum has an exhibit of how the engineering works. The importance of ecosystems in modern canals is also displayed.

One of the features of the museum is the ice well used to store ice and prevent it from melting. Ice houses date back thousands of years and were used by Alexander the Great. They pre-date modern refrigeration. Most ice wells are partially underground and surrounded by thick walls that insulate the ice and prevent it from melting. They are usually circular to resist pressure from the ground, and the lower surface area slows melting. A drain removes any excess water to keep the ice dry. Ice stored in ice houses lasted up to a year.

THE GRANT MUSEUM OF ZOOLOGY

21 University Street, London, WC1E 6DE
www.ucl.ac.uk/culture/grant-museum-zoology

As one of the UK's oldest history museums, the Grant Museum of Zoology might be small but it is home to 68,000 specimens. The museum was founded in 1827 by Robert Edward Grant as a teaching resource for the newly established University of London. One of his most famous students was Charles Darwin.

Highlights of the museum include a quagga skeleton, dodo bones, a rock python skeleton, and a jar of preserved moles. The Micrarium contains 2,300 microscope slides displayed in a lit area and shows the importance of microscopic creatures with a plethora of specimens. With rows of intriguing jars, there's much to take in on a short visit and no end of surprises within the displays. The museum offers free events, tours, and exhibitions.

ROYAL COLLEGE OF PHYSICIANS MUSEUM

Royal College of Physicians,
11 St Andrews Place, Regent's Park,
London, NW1 4LE
www.rcplondon.ac.uk

The Royal College of Physicians Museum was founded in the seventeenth century by William Harvey (1578–1657), the English doctor

Royal College of Physicians, London.
(Author, photographed with permission from the Royal College of Physicians)

and anatomist. Today, the collection of art and objects has expanded to include the history of doctoring and the achievements of prominent physicians. The first-floor gallery has portraits of physicians from history including Richard Mead, doctor to George II and Sir Isaac Newton. Within the museum are many medical objects including bloodletting bowls, a seventeenth-century surgeon's chest, and an impressive collection of apothecary jars. Outside, a delightful medicinal garden reflects the connection between plants and healing.

The Royal College of Physicians also offers guided tours of the heritage collections and gardens, with details on their website.

UNDERGROUND LONDON

THE POSTAL MUSEUM AND MAIL RAIL

15–20, Phoenix Place, London, WC1X 0DA
www.postalmuseum.org

The Postal Museum has an eclectic display of the history of mail in Britain. Exhibits include how technology changed how the post is delivered, and machinery used to sort the mail. However, one of the highlights of any visit is riding the Mail Rail underground train or walking a tunnel. Between 1927 and 2003, Royal Mail transported the mail underneath the streets of London, largely unseen by the public. The trains ran on a double 2ft gauge track along a tunnel 9ft wide. Over 6½ miles of tunnels covered the underground railway at its prime. Riding the Mail Rail train or walking the tunnels must be pre-booked and gives an excellent insight into the engineering that went into the design of the underground railway.

LEARNING OPPORTUNITIES

THE FRANCIS CRICK INSTITUTE

1 Midland Road, London, NW1 1AT
www.crick.ac.uk

The Francis Crick Institute is a biomedical organisation working across academia, industry and medicine to make scientific discoveries. The research includes an extensive cancer portfolio and vaccine studies. The Francis Crick Institute has free exhibitions and lectures and also actively encourages career development in science and technology through school visits and opportunities to 'meet a scientist'.

Francis Crick Institute. (Author)

ALAN TURING INSTITUTE

www.turing.ac.uk

The Alan Turing Institute is located within the British Library and is the UK's national institute for data science and artificial intelligence. A major project being led by the institute is the use of artificial intelligence to monitor London's air quality. Nine thousand Londoners die from the effects of poor air quality each year and the Alan Turing Institute, is using algorithms to predict pollution levels.

The Alan Turing Institute has scientific events including lectures and conferences throughout the year.

THE WELLCOME COLLECTION

183 Euston Road, London, NW1 2BE
www.wellcomecollection.org

Sir Henry Wellcome was an avid traveller and enthusiastic collector of books, objects, and art demonstrating the development of medicine. Somehow he managed to amass over 24,000 surgical instruments and between 5,000 to 10,000 glass jars, among other artefacts. The Wellcome Collection is a museum and exhibition centre with permanent and temporary displays. Highlights include exhibitions, and the free tours given by staff on the history of medicine. There are also lecture programmes at the Wellcome Collection and a cafe.

HOTELS WITH A CONNECTION TO SCIENCE

THE MEGARO HOTEL

Belgrove St, Kings Cross, London, WC1H 8AB
www.themegaro.co.uk

The brightly coloured building opposite King's Cross Station is a distinctive landmark, but the interior is impressive. Throughout the Megaro, the décor is inspired by the engineers and scientists who developed places like St Pancras Station and the railway network. Chemistry distilling jars are interspersed with pipework and laboratory decor. Magenta Restaurant has vibrant pink furnishings, and each bedroom has a distinctive style. The cocktail bar, Hokus Pokus, is inspired by local Victorian Dr James Morrison, who opened the British College of Health nearby in 1828. The cocktails have a distinctive alchemy theme.

Megaro Hotel, London. (Author)

4

EAST LONDON AND DOCKLANDS

London's East End is full of industrial heritage, and has been the destination of immigrants for centuries, many fleeing persecution. When the population of London rose between 1800 and 1900, so did poverty, social inequalities, and crime in the East End. However, the East End has a wider heritage story, including innovative industrial developments, medical research, and inspirational infrastructure. On the edge of the East End, Tower Bridge is one of the finest examples of bascule structures globally, while the neighbouring Tower of London was a prominent centre of scientific advances in the seventeenth and eighteenth centuries.

LONDON LANDMARK

THE TOWER OF LONDON

Tower of London, London, EC3N 4AB
www.hrp.org.uk/tower-of-london/visit

The Tower of London attracts thousands of people each year and is a must-see for any visitor. Built by William the Conqueror in the 1070s, this thousand-year-old building is full of history. From the execution of queens to the glittering Crown Jewels, there are many reasons why people visit the Tower of London.

Tips for Visiting

Pre-booking is advisable so you have a guaranteed entry time. The tower is best visited earlier in the day before the crowds arrive. Tower Hill tube station is close by, but visitors can also come by riverboat, getting excellent views of the city skyline from the Thames. Many prisoners arrived by boat, entering through the Traitor's Gate in medieval times. The Yeoman Warder tour times are on display just inside the main entrance. These tours are included in

Tower of London. (Andreyspb21 via Adobe)

the ticket price. They are a fascinating insight into the life of a beefeater in the Tower of London. Along with stories of imprisoned and executed people, the yeomen warders have several anecdotes about their unique lifestyles. Head to the Crown Jewels first, as this part of the Tower gets the most crowded, especially later in the day.

Scientific Highlights

The Tower of London has several sites connected with science.

The Trebuchet

Just inside the main entrance to the Tower of London is a recreation of a trebuchet like those used in medieval battles. Back in their day, the trebuchet was a brutal piece of armoury that inflicted significant damage to people and property, especially during sieges. Trebuchets work by using the energy of a falling counterweight to launch a projectile (known as the payload). Mechanical advantage is used to achieve a greater speed. The counterweight is balanced at a much shorter distance than the payload end, increasing its velocity. This enables the load to be launched at a higher speed to inflict as much damage as possible. The trebuchet was preferred over the catapult for this very reason. Trebuchets operate on an uneven balancing mechanism that can propel their load faster, whereas the balance on a seesaw, for example, is in the middle.

The White Tower

The White Tower is the oldest part of the complex and is the original royal palace lived in by kings and queens. The torture went on in the dungeons below, and the walls were so thick they were almost soundproofed. Three of the turrets are square, but one is round, and this was the site of the first Royal Observatory. Charles II appointed John Flamsteed as the first Astronomer Royal. He conducted experiments and observed the stars from the tower. At the time, there was a race among astronomers to map the stars. Flamsteed conducted several observations and was the first person to identify Uranus. Today this part of the White Tower is known as the Flamsteed Turret.

The White Tower was also home to the Board of Ordnance. Their primary role was in arms and weaponry, but they were also responsible for significant geographical surveys. The Board of Ordnance developed prolific map-making skills and was also responsible for teaching young cadets mathematical skills and draughtsmanship. The drawing rooms and offices were in the Tower of London before moving to Southampton in 1841. The department became the Ordnance Survey and Britain's national map-making organisation. The world's first national Geological Survey, the Ordnance Geological Survey, was established in 1835. It became the British Geological Survey in 1984 and is still active.

Crown Jewels

The crown jewels are a significant and popular attraction at the Tower of London. The diamonds within the crowns and sceptres are some of the largest in the world. Their origins are full of geological history. One of the most famous diamonds in the collection is the Cullinan, the largest diamond to be mined. It was discovered in South Africa in 1905 and was so large gemologists cut it into nine pieces. The largest gem formed the Star of Africa in the Sovereign's Sceptre. The second-largest piece is in the Imperial Crown. Scientists had been fascinated by the brilliance of the diamond for years, trying to comprehend how the Earth produced a gem of such clarity. Research by the Gemological Institute of America suggests the Cullinan was formed up to four times closer to the centre of the earth than other diamonds.

Diamonds are formed in the earth's mantle under high-pressure conditions. Seabed sediments containing carbon are dragged towards the centre of the earth. At this depth, rocks have molten metal, which fuses with the carbon forming giant diamond crystals. Volcanic eruptions carry rocks and gems back to the earth's surface, where they are mined.

Most diamonds form at 150km to 200km below the surface of the Earth, within the base of the continental tectonic plates. Some rarer and larger diamonds form much deeper and nearer where the mantle is slowly moving. The depths are estimated to be 350–700km beneath the earth. The chemical Bridgmanite was also found in the Cullinan. Bridgmanite is only found in very deep mantle areas, suggesting the diamond crystallised at these depths.

Ravens

Legend says that if the six ravens leave the Tower of London, England will be destroyed. As a precaution, their wings are clipped, and there are seven birds, so there is always a spare. The ravens even have their own raven master, a dedicated yeoman warder who cares for the birds on Tower Green. When John Flamsteed, the first Astronomer Royal, lived and worked in the White Tower he was continually irritated by the ravens as they swooped around his telescopes and messed on the instruments. It was either the ravens or Flamsteed who left, and no one would allow the birds to leave the Tower. Finally, Flamsteed departed for Greenwich, where Sir Christopher Wren and Robert Hooke built the current Royal Observatory.

Memorial to Alexander John Forsyth

The Reverend Alexander John Forsyth (1768–1843) was a Scottish Presbyterian minister with an interest in gunpowder. A keen gun enthusiast, Forsyth wanted to improve the unreliable flintlock method of loading guns. So he developed a way of using detonating chemicals to prime the gunpowder used in arms. The technique became known as percussion ignition and is still used today. In 1805 he was asked to work in the Tower of London under the Master General of Ordnance and conducted secretive scientific experiments into weaponry during the Napoleonic Wars.

There is a memorial to Alexander John Forsyth on Tower Green, close to the exit stairway.

Graffiti in the Salt Tower

In Tudor times, the Salt Tower was used as a prison, with some of the inmates carving graffiti into the walls. One prisoner, Hew Draper of Bristol, was charged with sorcery in 1561 and locked in the tower. He created an elaborate astrological clock on the walls, which can be seen today. Draper's fate is unknown.

The Royal Mint

The Royal Mint was located within the Tower of London and made the nation's money for over 500 years. During the reign of Edward I, there were significant problems with people clipping silver coins to melt down. As a result, the king brought coinage production into the Tower of London. But, over the next 500 years, kings and queens struggled to maintain the quality of the currency. Finally, Henry VIII instigated a debasement of the money by substituting cheap metals for silver. Eventually, Elizabeth I reinstated silver coinage, and the screw press was invented to make coins. Unfortunately, the working conditions exposed many workers to fatal doses of arsenic.

Isaac Newton was appointed Warden of the Royal Mint in 1696, and it was during this era he significantly improved accuracy and measurements. Newton saw an opportunity to apply scientific rigour to the work and, when he was appointed Master of the Mint three years later, took the opportunity to improve standards. His work included using his knowledge to ensure that the purity and weight of coins were equal. Newton also created the first currency of a unified Britain after the 1707 union with Scotland. He was concerned with the sheer number of forgeries and is known to have pursued several forgers through the courts. He also worked with the Mint's engravers, encouraging them to do private work to improve their expertise and make coins harder to forge. It resulted in significantly fewer forgeries passing into circulation. Isaac Newton became President of the Royal Society during his time as Master of the Mint.

Sir Walter Raleigh and the Bloody Tower

The Bloody Tower is infamous for the tragic story of the two princes who Richard III allegedly murdered in 1493. On three occasions, Sir Walter Raleigh was a prisoner in the Tower of London and had a cell in the Bloody Tower. His cell, which was significantly better than that of an ordinary prisoner, can be seen today. He conducted and wrote about scientific experiments during his stay and cultivated an exotic and medicinal garden outside the tower. The garden featured plants he had sourced on his travels. He created a herbal elixir from plants, known as Raleigh's herbal elixir, which contained rosemary, mint and bistort. In 1616 he was executed.

The Moat

In 2022, the moat surrounding the Tower of London was filled with wildflowers, creating a vibrant display and providing a habitat for bees in the city. Wildflowers will continue to provide biodiversity in the moat.

TOWER BRIDGE

Tower Bridge Rd, London SE1 2UP
www.towerbridge.org.uk

Tower Bridge was built between 1886 and 1894 and is one of the most famous London landmarks. Most people take a selfie, go on a river cruise, or walk across it, but only a few explore the

Tower Bridge. (Author)

engineering genius and how the bridge works. Many assume it is a drawbridge, but Tower Bridge is one of the world's finest examples of bascule bridges. The side spans are suspension bridges, while the centre span is 61m in length with twin bascules. Bascule is French for seesaw, reflecting the motion of the bridge as it opens and shuts. Each bascule weighs over 1,000 tons and moves around a pivot – and is lighter than a road to lift. The sides are equally balanced, resulting in less energy needed to open the bridge.

The bascules were originally operated by a hydraulic mechanism using steam to run the pumping engines, but have been powered by electricity since 1976. Energy is stored in accumulators, so it is available when needed. When the bascules open huge counterweights descend into the space beneath the bridge, contained by the concrete piers.

Visitors can walk across the bridge or sail under it. They can also access the glass walkway at the top, looking down on the bridge and river below.

One of the best ways to appreciate the engineering is to take a behind-the-scenes tour to view the original engines and bascules under the bridge and meet the engineers who operate the system. The tours operate between November and February when there are fewer bridge openings. They are led by one of the four drivers allowed to operate Tower Bridge, so they give an insight into the workings and engineering. The tour is not for the faint-hearted; a good fitness level is required to climb up and down many steps and access confined and often dirty spaces. The Victorian engine room is also included in the tour. Visitors must pre-book tours via the website, and they usually sell out.

Tower Bridge also holds concerts within the bascule chambers. These are listed on the Tower Bridge website.

ENGINEERING EXPERTISE

THAMES BARRIER PARK

N. Woolwich Rd, London E16 2HP

The Thames Flood Barrier protects 125 sq km of central London from flooding and stretches 525m across the River Thames. It is one of the largest movable flood barriers in the world. The Thames Barrier Park in the Royal Dockyard is one of the best places

Thames Barrier. (Author)

Green Dock at Thames Barrier Park. (Author)

in London to see the structure close up. However, the visitor centre and tours are operated from the Woolwich side of the Thames (see South-East London).

The garden is a 7-hectare space designed as a 'green dock'. The hedges are shaped like waves or the flood barrier itself and provide a microclimate within the park. The nearest underground station is Pontoon Dock.

CODY DOCK ROLLING BRIDGE

South Cres, London, E16 4TL
https://codydock.org.uk/cody-dock-rolling-bridge-update

A 12-tonne rolling bridge at Cody Dock opened in 2022 and operates innovatively using a hand crank. When the crank is turned, the structure inverts on tracks so shipping below can pass through. It is the opposite concept of a bascule, which uses hydraulics and a lot of energy. Instead, Cody Dock Bridge operates using balance and counterweights. The bridge allows the dock to be flooded so shipping can pass through, and it is also the link for walkers and cyclists on the River Lea footway. It also connects The Line Art Walk in East London. The nearest Tube station is Canning Town.

SPORTING HERITAGE AND SCIENCE

QUEEN ELIZABETH OLYMPIC PARK

The Queen Elizabeth Olympic Park is an urban space in East London and home to the sporting facilities used for the 2012 Olympic Games. Within the park are several places with scientific features accessible to the public.

ArcelorMittal Orbit

Queen Elizabeth Olympic Park,
5 Thornton St, London E20 2AD
https://arcelormittalorbit.com

Created by Anish Kapoor and Cecil Balmond for the 2012 London Olympic Games, the ArcelorMittal Orbit is an iconic structure in East London. It is a twisted looping structure that is the UK's largest sculpture and was constructed from steel, 60 per cent of which is recycled. In addition, 600 pre-fabricated star-like nodes have been engineered to create the famous structure. A slide created by Carsten Höller twists around

ArcelorMittal Orbit at Queen Elizabeth Olympic Park. (Author)

the design and is the world's longest tunnel slide. Visitors can see the iconic structure up close by travelling by lift to the viewing platform, where there are views across London. Steel mirrors distort shapes with reflective trickery.

For those who want to travel by slide to the bottom of the sculpture, there is a forty-second 178m ride that twists and turns like a corkscrew and is not for the faint-hearted. Speed is controlled via the helical shape, curvature, and slide gradient. Spiral stairs or a lift are the alternate ways down.

London Aquatics Centre

Queen Elizabeth Olympic Park,
5 Thornton St, London E20 2AD
www.londonaquaticscentre.org

The London Aquatics Centre has been designed for all levels of swimmer and has seen several world champions swim in the famous pool. The pool was designed to enhance the performance of athletes by reducing ripples and waves. The lane markers, gutters, and proportions are designed to minimise turbulence and improve performance. In addition, the pool has adjustable depths but is set to 3m deep for races. This leads to waves being deflected from the swimmer's movements, causing less turbulence in the pool. When the pool has fewer waves and ripples, swimmers move faster in the water.

The Velodrome

Queen Elizabeth Olympic Park,
5 Thornton St, London E20 2AD
www.better.org.uk/leisure-centre/lee-valley/velopark/faqs

You can learn four different biking sports at the Lee Valley Velo Centre, from BMX to racing. The iconic velodrome is a centrepiece and a focal point for geometry in cycling. The roof is a parabola, and the velodrome has been designed with the geometry of cycling in mind. However, the track is a symmetrical circuit that slopes at 42 degrees so cyclists can pick up speed using centripetal force. So although cyclists riding at speed on the banked sections look as though they are riding at an angle, they are in fact cycling upright, thanks to the effects of centrifugal force.

WALKING TOURS

THE LINE

0AX, Olympian Way, London W1D 1LP
https://the-line.org

The Line is London's first dedicated art walk and roughly follows the path of the Greenwich Meridian. It connects the three boroughs of Tower Hamlets, Newham, and Greeenwich. It runs between the Queen Elizabeth Park and the O2. The Line takes four hours to walk fully and is accessible to cyclists. It is partially wheelchair accessible, and an alternative map is downloadable from the website. Alternatively, you can walk sections as public transport is never far

The Line trail sign at Canning Town. (Author)

away. Some of the art installations have scientific links. These include:

Eva Rothschild's Living Spring. The stick-like installation is representative of nature.

Abigail Fallis. DNA DL90. This iconic structure is made of several shopping trolleys formed into a double helix, representing DNA. It was installed on the fiftieth anniversary of the discovery of DNA.

Rana Begum. Catching Colour. The colourful installation is representative of the reflections of light from the River Lea nearby.

Sculptures of shopping trolleys in a DNA helix by Abigail Fallis. DNA DL90. (Author)

Antony Gormley. Quantum Cloud. Gormley's intriguing installation in North Greenwich comprises 325 extended tetrahedral sections connected to over 3,500 of the same elements that extend into space. The art suggests a relationship between mass and energy and whether the body produces the field or vice versa.

LIGHTHOUSE HERITAGE IN EAST LONDON

TRINITY HOUSE

Trinity Square, London, EC3N 4DH
www.trinityhouse.co.uk

Trinity House is a charity concerned with the safety of shipping and the welfare of seafarers. It was incorporated by Royal Charter in 1514 and operates a fleet of lightships and the lighthouses around the coast of Britain. The tours give a comprehensive insight into the history and work of Trinity House. From a scientific perspective, you'll see several unique and fascinating items. The ground-floor entrance has a binnacle from HMS *Warspite* and also a 'curve of daylight' acetylene-regulating clock from the Farne Lighthouse used from 1910 to 1965. On the ground floor are terrestrial and celestial globes. Upstairs, the landing or quarterdeck leads to the Reading Room containing a sextant presented to Admiral Sir Richard Collinson, who searched for the ill-fated Franklin Expedition. Other objects include a clock and barometer from Shackleton's *Nimrod*. Trinity House also has original documents written by John Smeaton and many artworks relating to maritime history. Trinity House is open to the public for guided tours by pre-arrangement only and can tailor a tour to deal with specific aspects of history.

TRINITY BUOY WHARF

www.trinitybuoywharf.com

Located on the River Thames, Trinity Buoy Wharf is an arts and culture centre with significant scientific attractions. Many of the older buildings were used by Trinity House for its work with lighthouses and lightships and are now art studios and workspaces. Michael Faraday worked as a scientific adviser at Trinity Buoy Wharf for thirty years from 1836. A replica shed is one of London's smallest museums, depicting his workshop and containing information about his work on the Faraday Effect. Created by artists Ana Ospina and Fourth Wall Creations, the Faraday Effect is a delightful museum with sound effects and artefacts from Faraday's time in the area.

Trinity Buoy Wharf. (Author)

Faraday Shed at Trinity Buoy Wharf. (Author)

Trinity Buoy Lighthouse. (Author)

London's only lighthouse is at Trinity Buoy Wharf. It was built in 1864 and used as an experimental building to inform the design of lighthouses and lightships in Britain. For example, Michael Faraday used the experimental lighthouse to conduct his optical studies for maritime use. Among his inventions was a chimney which improved the problem of condensation from oil lamps and increased the amount of light produced. He also developed lenses used in maritime lighting in the 1840s.

The lighthouse is open at weekends and specific times listed on the website. Trinity Buoy Wharf is open from 7am–7pm daily, including the Faraday Effect Museum. On exiting Canning Town Underground Station, the site is signposted.

MUSEUMS WITH SCIENCE, MEDICAL AND ENGINEERING HERITAGE

ROYAL PHARMACEUTICAL SOCIETY MUSEUM

66-68 East Smithfield, London, E1W 1AW
www.rpharms.com/about-us/museum

This small museum is full of fascinating facts about how pharmacology developed. The displays portray the story of how the first apothecaries used herbal remedies and how prescribing and regulation emerged. There are dubious remedies, from polar bear grease for baldness to throat pills loaded with morphine. The museum sets out how antibiotics, pain relief and vaccines developed and the impact these have had on health.

HACKNEY MUSEUM

1 Reading Lane, London, E8 1GQ
https://hackney-museum.hackney.gov.uk

Hackney Museum is a small collection of items connected to local history. Of scientific interest is the section on the market gardening industry from the 1500s to 1850. In particular, the heritage of German nursery owner, Conrad Loddiges, is celebrated in the museum. Loddiges owned a business near Hackney Museum and grew orchids. Today, many of the trees at Abney Cemetery, one of the seven great cemeteries in London, came from Loddiges' horticultural work. Hackney Museum also has displays about mental health and healthcare in the area.

Hackney was also home to the first 'green' power station. The Vestry of Shoreditch opened a generator in 1846, which generated electricity by burning the area's rubbish. The generator provided hot water for the municipal bathhouse and sufficient power to light up Shoreditch.

Hackney Museum. (Author)

THE VAGINA MUSEUM

Arches, 275-276 Poyser St, London, E2 9RF
www.vaginamuseum.co.uk

The Vagina Museum is the first bricks and mortar museum dedicated to the female reproductive system and aims to break down taboos associated with women's health. In addition, the museum has exhibits about periods and other gynaecological issues.

MUSEUM OF LONDON DOCKLANDS

1 Warehouse, West India Quay, Hertsmere Rd, London, E14 4AL
www.museumoflondon.org.uk/museum-london-docklands

There are three floors of displays about the history and heritage of the London Docks in this free museum (special exhibitions have an additional charge).

Museum of London Docklands. (Author)

Scientific highlights include displays of why weights, measures, and loads play an essential part in dockside work. In addition, other collections include engineering and Brunel's work on the Thames Tunnel and the SS *Great Eastern*. Hydraulic power used in dock work, health and safety, and sampling ingredients are also exhibited. The London Docklands played a significant part in the Second World War, and an exhibit includes Maunsell Towers and the secret technology used in cables during wartime.

LEARNING OPPORTUNITIES

CENTRE OF THE CELL

Blizard Institute, 4 Newark Street, Whitechapel, London, E1 2AT
www.centreofthecell.org

Centre of the Cell is a science education centre based on London's Queen Mary University campus. The education centre is the first in the world to be located in a working biomedical lab so visitors can see scientists at work. Outside is an impressive sculpture known as the Neuron Pod. The site is designed to be interactive and playful to attract

Centre of the Cell, London. (Author)

younger people. During school hours, the space is reserved for schools to visit and engage in talks and workshops. Public sessions are booked via the website. Pre-booking is essential.

Further, along Newark Street, a statue commemorating Queen Alexandra is at the junction with the Royal London Hospital. The plinth has an engraving of people being treated with a Finsen Lamp for lupus. Queen Alexandra introduced the lamp to the hospital as an innovation project. Nearby, on Whitechapel Road almost opposite the Royal London Hospital, is 259 Whitechapel Road. It is here that Joseph Merrick, otherwise known as the Elephant Man, was first put on display to the public and discovered by Sir Frederick Treves.

SCIENTIFIC MEMORIALS AND STATUES

ALEXANDER PARKES PLAQUE

Berkshire Road, Wallis Road, London, E9

The first plastic in the world was made in Hackney in 1866. At the time it was named Parkesine, after its inventor, Alexander Parkes. A brown memorial plaque commemorates this scientific discovery in Wallis Road, Hackney.

As a sign of how science and society have moved on, a plastic-free shop is a short distance from the memorial and has inspirational ideas on how to be more eco-friendly. Refill Therapy is at 61-63 Wallis Road, Hackney Wick, E9 5LH. www.instagram.com/refilltherapy

At nearby Hackney Wick Station there are hexagonal glass pillars which form prisms when the sun is low. They are a nod to the local chemical dye industry. The underpass

Alexander Parkes sign in Hackney. (Author)

has a molecular chemical frieze in recognition of the chemicals industry that was once in the area.

THE COLOUR PURPLE

At one time, the colour purple was only worn by extremely wealthy people. However, aniline dyes were invented by accident in Cable Street, Shadwell, by teenager William Perkin in 1856. Perkin was experimenting at home and created the colour mauveine from aniline. He went on to establish a dye works in Greenford and laid the foundations of the organic chemical industry, revolutionising the fashion world, as everyone could wear purple. A memorial plaque to Sir William Henry Perkin (1838–1907) is outside his former home on Cable Street at the junction with King David Lane.

THE WHITECHAPEL FATBERG MANHOLE COVER

Court Street and Whitechapel Road, London, E1 1DG

In 2017, a 130-ton mass of oils, wet wipes, grease, and general rubbish clogged up the city sewer system in Whitechapel. As a result of the solids being unable to dissolve, engineers found one of the largest fatbergs ever discovered in the city, stretching over 820ft in length. It took sewer engineers over three months to remove the fatberg, and a section was dried and donated to the Museum of London. The commemorative manhole cover opposite the Royal London Hospital is a salutary reminder of solubility.

JAMES PARKINSON BLUE PLAQUE, SHOREDITCH

1 Hoxton Square, Shoreditch, N1 6NU

James Parkinson was born on 11 April 1755 and studied at the London Hospital Medical College, where he qualified as a surgeon. He is most famous for publishing *An Essay on the Shaking Palsy*, in 1817, which established Parkinson's as a medical condition. James Parkinson lived and worked at 1 Hoxton Square for most of his life. He was also a social reformer and prominent geologist. Parkinson published *Organic Remains of a Former World* in 1804, which was the first scientific account of fossils.

SOUTH-EAST LONDON

South-East London has significant connections to science and medicine. Elephant and Castle is the birthplace of Michael Faraday, one of Britain's most celebrated scientists. South-East London is where Marc Brunel built the Thames Tunnel and where Joseph Bazalgette constructed Crossness Pumping Station. It is also one of the most deprived areas of London, with stories of several pioneering people who worked to reduce health and social inequality.

LONDON LANDMARK

THE SHARD

32 London Bridge St, London, SE1 9SG
www.the-shard.com

The Shard is one of the tallest buildings in Western Europe at 310m. It was designed by Renzo Piano and opened in 2012. Its iconic sloping shape has several architectural and scientific features. Pioneering engineering methods were used to construct the Shard, including top-down construction where foundations are dug as the core is built upwards. The process saves time; it was the first time it had been used on a core. Steel columns were installed inside the concrete piles first, so most of the ground floor was built. Next, they dug downwards, creating the basement. The steel columns supported the ground flooring, enabling the engineers to build up simultaneously.

The main framework of the building consists of horizontal beams making up ceilings, floors and roofs. They hold up beams and form walls. In skyscraper construction, the weight of all the materials is calculated to ensure the steel or concrete beams will not get crushed. Engineers also check the forces acting on beams to ensure they will not break due to force compression or force tension. Steel is used in construction because it is more robust than wrought iron due to the added carbon atoms. The structural steel in the Shard weighs around 12,500 metric tonnes, and the core is made of concrete, enabling the building to withstand wind forces. Concrete is also

The Shard. (Author)

used in room construction as it absorbs noise. The spire is made from steel, with the beams and columns bolted and welded together.

Visiting the Shard

Many businesses are located in the Shard. However, visitors can experience the skyscraper in several ways. The viewing gallery on the seventy-second floor has a panoramic view higher than many other platforms in London and offers views of up to 40 miles on a clear day. Visits to the View From the Shard should be pre-booked.

The Shangri-La Hotel is a luxury hotel on Level 34 where guests can enjoy views from a height of 125m. The hotel has 202 rooms, three dining areas, and London's highest infinity swimming pool.

The Shard has several restaurants open to the public. The Gong Bar on level 52 is the highest hotel bar in Europe. Aqua Shard is on Level 31 with panoramic views across London and serves contemporary British cuisine and afternoon tea.

ENGINEERING EXPERTISE

CROSSNESS PUMPING STATION, ABBEY WOOD

Bazalgette Way, Abbey Wood, London, SE2 9AQ
www.crossness.org.uk

By the early Victorian age, London had a significant issue with sewage, which was literally being pumped into the River Thames. In 1831 London experienced a deadly cholera outbreak, with a further epidemic in 1848, which killed hundreds of people. In 1858, the River Thames was overflowing with sewage, commonly known as the Great Stink, because of the smell infiltrating central London, including the Houses of Parliament. An increasing population and the lack of a sewerage system in London had resulted in a major public health hazard. Parliament realised something would need to be done to resolve the issue and commissioned Sir Joseph Bazalgette to construct a solution. As a result, Bazalgette created the London Embankment and designed a network of massive tunnels that carried sewage eastwards to Crossness on the South Bank of the Thames and Barking on the north side. Once the sewage arrived at Crossness, it was held in a closed reservoir until released into the River Thames at low tide, where it was taken away towards the North Sea and dispersed. Bazalgette had the foresight to plan for an increase in London's population and to test the system against significant rainfall.

In 1865, Bazalgette designed an ornate sewage pumping station at Crossness that is open to the public. It is an outstanding example of Victorian engineering and a masterpiece in architectural design.

Getting there

The easiest and fastest way to get to Abbey Wood Station from central London is via the Elizabeth Line. Abbey Wood is also served by National Rail. From Abbey Wood Station, there is a long thirty- to

Crossness Sewage Treatment Works. (Steve via Adobe)

forty-minute walk to the pumping station, or Abbey Wood Minicabs runs a taxi service to the site.

Tours

There are three ways to visit Crossness Pumping Station, and all must be pre-booked. There are designated open days, guided tours and photography visits. The website lists days when the pumping station is open.

Highlights of Crossness Pumping Station

There are several areas of the pumping station to experience on a guided tour of Crossness.

The Exhibition

Behind the ticket desk is a comprehensive exhibition outlining the history of sewage systems in London, the role of Joseph Bazalgette, and the connection to public health. The exhibition highlights the importance of adequate drainage and also has a collection of historic toilets (The Toilet Timeline).

The Octagon and Pumps

The highlight and iconic view of Crossness Pumping Station is the ornate Octagon and pumping engines. The Octagon was initially designed to let light into the building, as the original roof had small dormer windows. It was ornately decorated

Crossness Pumping Station. (Roger de Montfort via Adobe)

before falling into disrepair when the building was decommissioned. Volunteers worked on the rusty ironwork and cleaned it up, restoring it to how it would have looked in its Victorian heyday. Today, the Octagon is an exquisite feature of the pumping station in red and green. A closer look at the latticework reveals intricate detail in the metal, such as decorative figs and senna pods as a Victorian nod to bowel functions and sewage.

At the height of its use, four steam-powered rotative engines pumped the sewage at Crossness, allowing it to be separated correctly and filtered into the Thames. James Watt built the engines, each providing 125hp with 43ft iron beams and massive flywheels weighing over 50 tons. Each engine was given a suitably Victorian name from the main members of the royal family: Victoria, Albert Edward, Prince Consort, and Alexandra. The four engines continued pumping, using low steam produced by Cornish engines until 1895, when an upgrade was needed. A new engine house was built housing two triple-expansion engines, which helped with pumping while the beam engines were upgraded. However, by the 1930s, the engines were working a lot less; by the 1950s, work had ceased.

Sewage disposal into the Thames was found to have tragic circumstances in 1878 when the *Princess Alice* steamship sank following a collision in the Thames near Crossness. More than 650 people died, and the high death rate was believed to be due to the 75 million gallons of sewage pumped into the Thames twice daily. As a result of the tragedy, a Royal Commission recommended solids should be separated from sewage and sedimentation tanks were added to the site in 1891.

Today, Prince Consort runs using steam and operates on open days only. Victoria is currently being restored by the volunteers. Visitors on guided tours can get to the upper level of the pump house to see the engines and architecture up close.

The Valve House

Several small engines are displayed in the Valve House, and many are undergoing restoration by volunteers. There is also a small display of equipment used by Thames Water to measure tides and water leaks.

The Volunteers

Crossness is run by volunteers, from running the cafe to restoring engines, and they have a wealth of information between them. Their enthusiasm and energy shine through. You'll find people at work restoring items all over the complex and conducting the tours.

KIRKALDY TESTING MUSEUM

99 Southwark Street, London, SE1 0IF
www.testingworks.org.uk

A quote: 'Facts, not opinions', is carved over the doorway of the Kirkaldy Testing Station in Southwark. Entering is like stepping back in time to the Victorian era. The smell of grease on carefully restored machinery, the uneven original floorboards, and the architecture are all reminders of the importance of this building.

The Industrial Revolution saw engineering developments transform the world, but engineers were on a

Kirkaldy Testing Station. (Author)

crash course, learning what worked and what was unsuccessful. Finally, one man, David Kirkaldy, understood the importance of testing materials. He spent thirteen years developing testing machines before publishing his findings and opening the station on Southwark Street in 1862. The 47ft 7in Universal Testing Machine designed, patented and commissioned by David Kirkaldy can be seen in the museum today. The Universal Testing Machine was the only one of its kind in the world for years. It works horizontally, exerting loads of up to 440 tons on the specimen being tested with a water-hydraulic cylinder and ram.

Other machines in the testing station include Brinell and Vickers hardness testing machines, Charpy and Izod Pendulum Testers, Avery tension machines, and Riehlé tension and compression machines. David Kirkaldy kept diligent records of all his tests and results, including a museum of fractures showing the failures. Three generations of Kirkaldys ran the testing station until 1965.

Within the testing station, tests were performed on many of London's building projects, the Comet airliner and the Forth Road bridge. Kirkaldy's work paved the way for the health

Kirkaldy Testing Station. 'Facts not ideas' was a quote from David Kirkaldy. (Author)

and safety movement, which is a critical factor in science, medicine, and engineering.

THE BRUNEL MUSEUM

Railway Avenue, Rotherhithe, London, SE16 4LF
https://thebrunelmuseum.com

You can literally walk in the footsteps of the Brunel family at the Brunel Museum in Rotherhithe. The tiny museum is where Marc Brunel devised his ambitious plans for the Thames Tunnel in 1825 and where work began. Brunel's Thames Tunnel was the first of its kind under a navigable river in the world. Incredibly, Marc Brunel was inspired by a worm when planning the tunnel design.

Teredo navalis, or the shipworm, has sharp horns that grind wood into powder as it bores through ships. As it bores it eats and excretes mud, which lines the tunnel. An inspired Brunel decided to copy the shipworm to bore a tunnel and he achieved this by devising a tunnelling shield. The massive wooden grid contained thirty-six cells, with one worker standing in each cell, digging the

Brunel Museum. (Author)

mud in front of them. Once 11.4cm had been removed, the shield moved forward and the process recommenced. It was hard, laborious work, but the tunnel was completed. Unfortunately, the tunnel flooded several times but finally opened in 1843. The excavation tool concept designed by Marc Brunel to dig the Thames Tunnel was also used to dig the Channel Tunnel in the twentieth century.

There is an interesting display of plans and artefacts belonging to Marc and Isambard Kingdom Brunel, however, a tour of the tunnel entrance is a highlight. Guided tours enter the shaft, descending stairs to a platform. Ahead is the entrance to the tunnel, where you'll hear the story of how it was constructed and the challenges faced by the Brunels. Look out for the small doorway at the top of the shaft. It's where Isambard Kingdom Brunel escaped after the tunnel flooded in 1828. The Thames Tunnel was adapted as a foot tunnel, and part of it is used by the London Overground today. However, it never fully functioned as an entire transportation route.

Brunel Museum. (Author)

MEDICINE AND HEALING

THE OLD OPERATING THEATRE MUSEUM AND HERB GARRETT

9a St Thomas St, London, SE1 9RY
https://oldoperatingtheatre.com

Accessed by a steep, narrow spiral stairway, The Old Operating Theatre Museum and Herb Garrett are housed in the 320-year-old church building of St Thomas Hospital. The Herb Garrett was used as a herb store for medicinal supplies, and the historic operating theatre was added in 1822 and is the oldest surviving operating theatre in Europe. Today, the museum features displays of herbal remedies used by apothecaries and strange concoctions such as Dr Mead's Famous Snail Water (apparently a cure for syphilis). There is also a range of surgical instruments, including several used in women's

Old Operating Theatre. (Author. Taken with permission from the Old Operating Theatre Museum and Herb Garret)

surgery, which took place in the adjacent theatre. In addition, visitors can explore the historic operating theatre, which has steep and narrow steps. It was initially used as a theatre for women's surgery. Access to the building is limited for people with disabilities because of its historic nature.

TALKING STATUES AT GUY'S HOSPITAL

Guy's Hospital Entrance, St Thomas St, London, SE1 9RT
www.talkingstatueslondon.co.uk/statues.php

Talking statues connected to medicine are within the historic entrance to Guy's Hospital. The figure of Asclepius is positioned above the main entrance. Asclepius was the son of Apollo and is believed to have been a skilled physician who practised in Athens in 1200 BC. Today, the Rod of Asclepius represents the medical profession worldwide. By using a QR scanner, you can hear the statue speak.

Within the courtyard at Guy's is another talking statue. John Keats was born in London in 1795 and became a surgeon apothecary at Guy's

John Keats talking statue at Guy's Hospital. (Author)

Hospital before focusing on poetry. Some of his more famous poems include 'To Autumn' and 'Bright Star'. Unfortunately, he died of tuberculosis at the age of just 25. By using the QR code you can make him talk.

Thomas Guy, the founder of Guy's Hospital, is another talking statue in this area. He founded the hospital to treat the destitute in 1722.

FLORENCE NIGHTINGALE MUSEUM

St Thomas' Hospital, 2 Lambeth Palace Road, London, SE1 7EW
www.florence-nightingale.co.uk

One of the most famous women in the nineteenth century, Florence Nightingale, is best known for her influence on nursing. Her legendary work in the Crimean War raised standards in nursing care and the museum dedicated to her life has displays about her work with wounded soldiers. However, Florence had many more attributes and her mathematical skills also saved lives. She was a statistician, using a rose diagram to demonstrate death rates, linking it to good hygiene and nursing care. The museum displays her statistical research and the impact she also had on prison reform in England. Other exhibits show aspects of her early life, including her belief that she had a calling from God to care for the sick. Her legacy as a nurse is also featured in this captivating museum.

MARY SEACOLE STATUE

St Thomas' Hospital, Westminster Bridge, The Queen's Walk, London, SE1 7GA
www.maryseacoletrust.org.uk

A statue dedicated to Mary Seacole is within the grounds of St Thomas Hospital, near the main entrance. Mary was a courageous lady who nursed soldiers on the battlefields of Crimea. However, for years her contribution to nursing was not accepted by the authorities, mainly due to her ethnic background. Martin Jennings designed the statue with a disc representing the world and her travels. Her figure shows her marching defiantly with her bag of potions.

Mary Seacole statue at St Thomas Hospital. (Author)

SCIENCE GALLERY LONDON

Great Maze Pond, London, SE1 9GU
https://london.sciencegallery.com

Science Gallery London is a public space within King's College. The gallery holds science-related exhibitions throughout the year and opportunities for younger people to engage with the scientific community.

THE GORDON MUSEUM OF PATHOLOGY

Hodgkin Building, Guy's Campus, London, SE1 1UL
www.kcl.ac.uk/lsm/centre-for-education/museums/gordon-museum

The Gordon Museum of Pathology is the largest medical museum in the United Kingdom. It is not open to the general public but medical professionals and research students can visit by prior arrangement with the curator. The museum contains several unique artefacts, including Lister's antiseptic spray and the original adrenal glands that inspired Thomas Addison to describe the medical condition named after him. The Joseph Towne anatomical and dermatological wax models are within the museum, collections of specimens that belonged to Astley Cooper and Richard Bright, and the Lam Qua pre-operative tumour paintings.

BETHLEM MUSEUM OF THE MIND

Bethlem Royal Hospital, Monks Orchard Rd, Beckenham, BR3 3BX
https://museumofthemind.org.uk

The Bethlem Museum of the Mind is a free museum celebrating people with mental health problems. It is located within the site of the Bethlem Royal Hospital, which was established in 1247 and was the first institution in the United Kingdom to specialise in mental ill health. The hospital has been located in South London since 1930.

The museum has a remarkable collection of historical items, artwork and archives, including over 1,000 pieces of art. There are also temporary exhibitions and special events.

MUSEUMS WITH SCIENCE, MEDICAL AND ENGINEERING HERITAGE

EXPERIENCE THE FARADAY CAGE AT SOUTHWARK HERITAGE CENTRE AND WALWORTH LIBRARY

147 Walworth Road, London, SE17 1FZ
https://heritage.southwark.gov.uk

Southwark Heritage Centre is a treasure trove of local history and exciting exhibitions. One of the prominent residents celebrated by the museum is Michael Faraday, who was born in nearby Newington Butts. Southwark Heritage centre has a Faraday cage where visitors can experience the effects of electromagnetism. The room is reached by going through the library and up the stairs. To experience the phenomenon, book some time through the library. Unfortunately, visitors cannot use the internet in the meeting room due to the Faraday effect.

The Cuming Collection in the museum has other items relating to Faraday, including a disk dynamo from the 1851 Great Exhibition and a family Bible. Richard Cuming (1777–1870) lived on Walworth Road and was an avid collector of things including fossils. On his death, his son Henry passed the collection to the Borough of Southwark, where they form a significant part of the Heritage Centre. Other science-based items in the collection include medical instruments.

SCIENCE ON THE STREET

THE LONDON BRIDGE ORGAN

Stainer Street, London, SE1
www.pipe-up.org.uk/london-bridge-organ

London Bridge Station has an organ that can be used by any member of the public. It is used by renowned organists and people who want to try it out. Organs have a row of pipes that sit on a wind chest. Bellows push air into the chest and through the pipes, causing them to oscillate and produce a sound. Sounds are also controlled by the shape and design of the pipes.

MILLENNIUM BRIDGE

Thames Embankment, London EC4V 3QH

The Millennium Bridge is a steel pedestrian suspension bridge that traverses the River Thames from Upper Thames Street in the City of London to

Millennium Bridge. (james-padolsey unsplash)

Bankside and was originally opened in 2000. However, the crossing earned the name the 'Wobbly Bridge' because of the way it swayed when it was first installed. This led to its closing for two years. The cause of the swaying was attributed to synchronous lateral excitation. This phenomenon is due to the number of people naturally swaying as they cross the bridge in step, causing it to oscillate. In turn, the people travelling sway in time to the oscillations, causing the bridge to wobble. Researchers investigated the phenomenon as a result of the swaying. Initially, they thought that strengthening the bridge would improve it. However, it would significantly change its appearance. So instead, dampers were fitted to the bridge to dissipate energy from the movement of people. There has been no significant problem since; however, the bridge has still retained its reputation as a wobbly structure.

BLUE PANDEMIC POSTBOX

Lambeth Palace Rd, London, SE1 7EW

Postboxes are usually red in the United Kingdom. However, during the Covid-19 pandemic, five pillar boxes were painted blue in recognition of the work done by the NHS. One of these boxes is on Lambeth Palace Road, near the entrance to the Florence Nightingale Museum.

NATIONAL COVID MEMORIAL WALL

South Bank, London

The National Covid Memorial Wall was started by volunteers in April 2021. It is a mural of painted hearts where the public can write something about a loved one who died during the Covid-19 pandemic. New hearts appear as the old ones are used up. The memorial wall stretches for 500m and is opposite the Palace of Westminster.

COADE STONE LION

Westminster Bridge

The magnificent lion sculpture at the southern end of Westminster Bridge is made of Coade stone. Eleanor Coade (1733–1821) developed the artificial stone used in neoclassical buildings. *Lithodipyra* is a weather-resistant, durably moulded ceramic stoneware used on buildings and in garden ornaments. Eleanor perfected the clay recipe and the firing of the stone.

National Covid Memorial. (Author)

Coade stone lion on Westminster Bridge. (Author)

Coade stone was used in buildings such as the Royal Hospital, Greenwich, and Buckingham Palace.

FARADAY MEMORIAL

Elephant and Castle Roundabout, London, SE1 6TG

The massive stainless steel structure on the Elephant and Castle Roundabout looks like an electricity substation. It is a brutalist-style memorial to Michael Faraday, who was born nearby in Newington Butts on 22 September 1791. However, it also acts as an electricity substation for the Northern and Bakerloo Lines. The memorial was designed by Rodney Gordon in 1959 and completed in 1961. A stone plaque on the memorial site reads: 'This stainless steel sculpture commemorates Michael Faraday (1791–1867) English Chemist and physicist known for his research into electricity and magnetism who lived locally.'

Faraday Memorial, Elephant and Castle. (Author)

DR SALTER'S DAYDREAM

Bermondsey Wall West, London, SE16

Adjacent to the Angel Pub and a short walk from the Brunel Museum are the remarkable four statues making up Dr Salter's Daydream. The sculptures were created by Diane Galvin and installed in 1991. A bespectacled man can be seen waving to a little girl. His wife Ada is nearby, and a cat sits on a wall overlooking the Thames.

Dr Alfred Salter was born in 1873 and studied medicine at Guy's Hospital. After qualifying as a doctor, he came to Bermondsey but was shocked by the poverty he encountered. Dr Salter opened a clinic that was free to working-class

Dr Salter's Daydream. (Author)

people. He married his wife Ada in 1900, and together they worked to improve health conditions for the people of Bermondsey. Ada was elected the first female Mayor of Bermondsey in 1922. The couple dedicated their lives to improving public health and lived in the area instead of a more wealthy neighbourhood. However, they made a terrible sacrifice. Their daughter Joyce died age just 8 years old from scarlet fever in 1910.

Dr Salter installed a solarium for TB sufferers and showed educational films from vans in addition to providing free healthcare. As a result, by 1935, the infant mortality rate had dropped from 150 to 69 a year, and no mother died in childbirth in Bermondsey.

Dr Salter's Daydream showing his daughter Joyce. (Author)

THE DNA DOORS

King's College Waterloo Campus, 150 Stamford Street, London, SE1 9NH

The doors to King's College on Stamford Street have been designed to represent a DNA double helix as you walk through them. These scale models embedded in the revolving doors educate people about DNA and its structure. The discovery of the DNA molecule in 1953 was based on work at King's College London by Dr Rosalind Franklin and Professor Maurice Wilkins.

DNA doors at King's College, London. (Author)

BOTANICAL AND NATURAL HISTORY HERITAGE

THE HORNIMAN MUSEUM AND GARDENS

100 London Road, Forest Hill, London, SE23 3PQ
www.horniman.ac.uk

The Horniman Museum is a vast collection of treasures amassed mainly by one man. It originally opened as part of the Horniman family residence in 1890 and is named after Frederick Horniman, who ran the family tea business. Horniman later became an MP. He was known as a social reformer and strove to improve welfare for people in South London. Horniman's collection became the museum we know today.

The Horniman Museum has several scientific features. There is an extensive natural history and anthropology section. One of the most famous displays is the giant taxidermy walrus exhibit. A vast collection of insects with over 4,700 butterflies and over 3,000 plant specimens are other highlights. Glass cases contain a range of stuffed animals, including a dodo and ostriches. In addition, children will find several educational and interactive displays to enjoy.

In the gardens, the Butterfly House is home to many moths and butterflies, and the gardens have several trees that were planted more than 100 years ago. There is also a small aquarium. The Horniman Museum also has several ongoing projects studying natural history.

GERALDINE MARY HARMSWORTH PARK

Kennington Road, SE1
www.southwark.gov.uk/assets/attach/2278/Geraldine-Mary-Harmsworth-parks_trails.pdf

The Geraldine Mary Harmsworth Park is on the site of the old Bedlam Hospital and the Imperial War Museum. Within the public park is an Ice Age Tree Trail that outlines the trees that began to grow in Britain during the 5,000 years of ice retreat. There are thirty-four trees indigenous to the United Kingdom, and each is marked with a description and number corresponding to the tree's chronological order in colonising Britain.

Horniman Museum Glasshouse. (Author)

THE GARDEN MUSEUM

Lambeth Palace Road, London, SE1 7LB
https://gardenmuseum.org.uk

Located in the old St Mary's Church adjacent to Lambeth Palace, the Garden Museum has an exhibition of the history of gardening, including the early plant hunters and horticultural development. John Tradescant (1570–1638) was the first great gardener and plant hunter and lived in the Lambeth area. He travelled to the Arctic Circle in search of plants. His son John (1608–12) was married in St Mary's Church, Lambeth, and travelled to America in search of plants. He succeeded his father as Royal Gardener and started a botanical garden in Lambeth.

The Sackler Courtyard Garden was designed by Dan Pearson and was inspired by John Tradescant's journeys collecting plants. Among the foliage are the ornate tombs of the Tradescants and also the tomb of Captain William Bligh, captain of the *Bounty* famed for its mutiny.

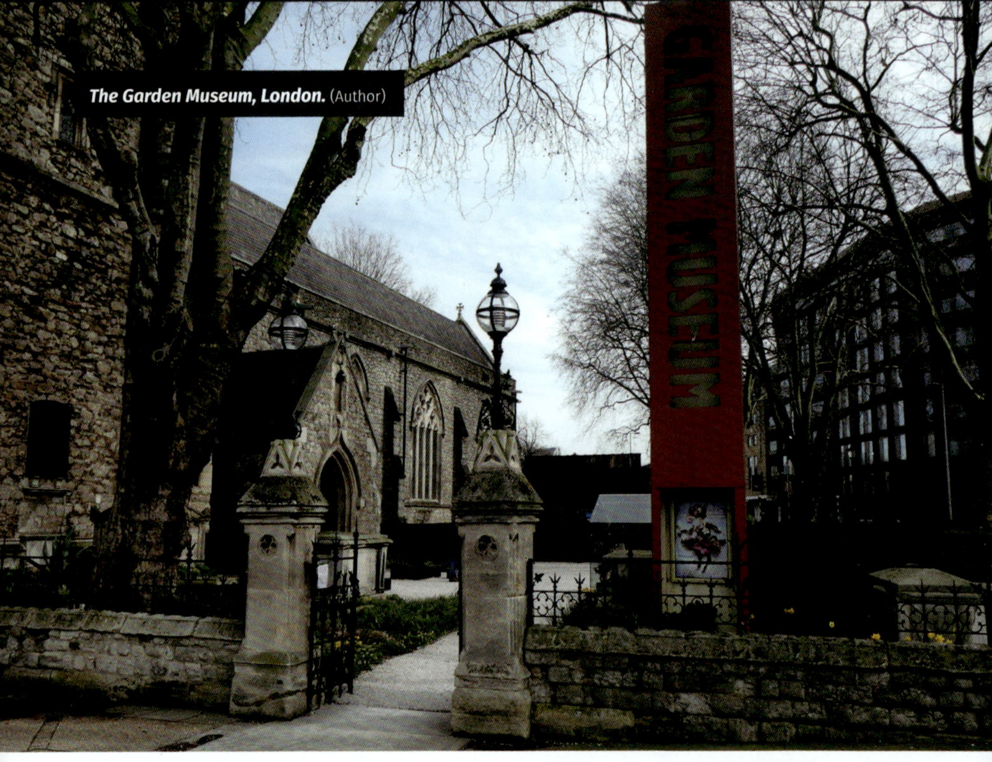

The Garden Museum, London. (Author)

BERNIE SPAIN GARDENS

South Bank, London
https://southbanklondon.com/attractions/bernie-spain-gardens

Bernadette Spain (1936–84) trained as a clinical psychologist and was relentless in her campaigns to reduce health inequalities and improve housing in South London. Her work with the Coin Street Action Group changed the area's landscape. The Bernie Spain Gardens are a delightful green space between the Oxo Tower and Gabriel's Wharf on the South Bank.

MORDEN HALL PARK

Morden Hall Road, Morden, London, SM4 5JD
www.nationaltrust.org.uk/morden-hall-park

Morden Hall Park is owned by the National Trust and is a 125-acre site of woodland and gardens to explore. The River Wandle flows through the park, which once depended on the snuff mills for its wealth. Today, visitors can see what life in a mill was like, explore the rose gardens, and enjoy the cafe. There is also a reverse

Bernie Spain Gardens.
(Author)

Archimedes Screw, one of the earliest hydraulic machine designs. It is named after the Greek mathematician Archimedes, who first described the principle in 234 BC. Water is pumped using a screw-like device inside a tube from one end to another, creating energy.

RESTAURANTS, PUBS AND BARS CONNECTED TO SCIENCE

THE WATCHHOUSE CAFE, ROTHERHITHE

69 Saint Marychurch St, London, SE16 4HZ
https://twitter.com/watchhousecafer

In the eighteenth and early nineteenth centuries, churchyards had watchhouses and towers where parish members guarded graves. The Watchhouse Cafe is located in the historic building where people once protected the neighbouring churchyard from the body snatchers who sold corpses to Guy's Hospital. Today, you can get good coffee and snacks in this delightful cafe.

THE MAYFLOWER

117 Rotherhithe Street, Rotherhithe, SE16 4NF
www.mayflowerpub.co.uk

The Mayflower lays claim to being the oldest pub on the River Thames, dating from 1550. It is most closely connected with the Pilgrim Fathers and was the original mooring point for the *Mayflower* in 1620. It was also used as a meeting place by the Royal Society to celebrate Marc Brunel's birthday, form the Tunnel Club, and lobby for Government funds.

Mayflower pub, Rotherhithe. (Author)

GREENWICH

Famous for maritime history, Greenwich has long been associated with scientists and engineers. Located on the famous Meridian Line, Greenwich is renowned for astronomy and timekeeping, with a rich heritage of museums and art. It is home to the Royal Observatory, where the science of the stars and navigation was studied. Maritime history and navigation are at the heart of Greenwich and have contributed significantly to scientific developments.

Getting to Greenwich

One of the best ways of travelling to Greenwich is by boat, as you'll see many London attractions along the route and will arrive on the waterfront with views of the Old Naval College and *Cutty Sark*. Boats operated by Thames Clippers travel regularly from central London to Greenwich. The Docklands Light Railway also has stops at *Cutty Sark* and Greenwich.

Tips for Visiting

Many of the historical sites are within a short distance of each other. The Royal Observatory and *Cutty Sark* have a combination of ticket offers, saving money on fees. The Old Royal Naval College entrance fee allows re-entry for up to a year, with timed tickets in operation for guided tours and the Painted Hall. Most venues operate a pre-booking system, especially for bespoke tours and events.

LONDON LANDMARK

ROYAL OBSERVATORY, GREENWICH

Blackheath Ave, London, SE10 8XJ
www.rmg.co.uk

The Royal Observatory at Greenwich is one of London's most prominent scientific heritage buildings. It is located in Greenwich Park and is most notably home to the Meridian Line and Greenwich Mean Time. When Charles II appointed John Flamsteed as the first Astronomer Royal in 1675, he wanted to see developments in understanding the

Royal Observatory, Greenwich. (Lefteris Papaulakis via Adobe)

stars and heavens and precision around longitude, especially with Britain becoming a dominant sea power. In 1884 that work culminated in the world setting its clocks to the time at the Greenwich meridian or zero longitude.

The Royal Observatory was home to the Astronomers Royal and comprised several historic buildings, including Flamsteed House, Great Equatorial Buildings, and Meridian Building. The Altazimuth Pavilion and South Building are now the Astronomy Centre and Peter Harrison Planetarium.

Standards of Length

A set of standard lengths was introduced in Parliament in 1760. In 1824, the Weights and Measures Act redefined standard measures in England. In 1838 the Astronomer Royal, George Biddell Airy, chaired a committee to reconstruct the standard national in weights and measures after they were destroyed when the Houses of Parliament burnt down in 1834. A plaque indicating the British Imperial Standards of Length was installed outside the gates of the Royal Observatory in 1859. The plaque both informed the public and was a visible indicator of the Royal Observatory's role in moderating technical and scientific standards.

Flamsteed House

The original Observatory was located in Flamsteed House and built in 1675–76

Standards of Time at the Royal Observatory, Greenwich. (Coward Lion via Adobe)

by Sir Christopher Wren and Robert Hooke. Its location meant that, at the time, it had a clear view of the skies, away from the smoke and pollution in central London.

On the ground floor of Flamsteed House was the living accommodation for the Astronomers Royal and their families. Today, the rooms have exhibitions about the lives of Flamsteed and another Astronomer Royal, Nevil Maskelyne. On the upper floor is the Octagon Room, where telescopes were housed to observe the skies.

Meridian and Great Equatorial Buildings

The Meridian building was originally a shed where John Flamsteed kept his transit quadrant mounted on a wall in the first attempt to define the Greenwich Meridian line. Over time, subsequent meridians moved eastwards, culminating in the work of George Biddell Airy, who became Astronomer Royal in 1835 and designed Airy's Transit Circle Telescope, which defined the Greenwich Meridian Line. The Great Equatorial Building was originally a telescope tower. It was added in 1857. Today, the building houses the 1894 refracting telescope, which is the eighth largest in the world. The iconic onion dome roof can be seen from Greenwich Park, but getting inside it to see the Great Equatorial Telescope is one of the best experiences in the Royal Observatory.

Time and Longitude Gallery

There are several famous clocks and instruments in the Time and Longitude Gallery. The Harrison Clocks are a significant feature, with H1, H2, H3 and H4 displayed with their detailed and exquisite designs. John Harrison's clocks enabled sailors to measure longitude at sea. H1 was developed in 1735, and Harrison tested it at sea before refining successive clocks and instruments. His work drastically reduced the number of shipwrecks, improving safety. Over Harrison's lifetime, the clocks were adjusted for temperature and ran without lubrication. Flamsteed's quadrant is also displayed in this gallery.

The Altazimuth Pavilion and South Building

The Altazimuth Pavilion was built in the late nineteenth century to house more telescopes. Today, it is a public observatory with computer-controlled systems to adjust for light pollution so the night skies can be viewed clearly. The Annie Maunder Astrographic Telescope (named after one of the first women to work at the Royal Observatory) has been one of the key developments in enabling the Observatory to open to the public. The building also has an auditorium housing the Peter Harrison Planetarium. Access to shows at the Planetarium must be pre-booked.

The Meridian Line

One of the most popular places in the Royal Observatory is the metal strip marking the Prime Meridian Line, which centres on Greenwich. Longitude was a significant problem until John Harrison devised a series of portable clocks in 1765, which were accurate to within a few seconds. The rotation of the Earth's position relative to the sun is linked to a relationship between time and longitude. Although people kept time by looking at sundials and church clocks prior to 1840, timekeeping at sea was critical to navigation.

Today, visitors can stand across the Prime Meridian Line with a foot in both hemispheres and take an iconic selfie. Within the courtyard are some of the best views across the London skyline that are uninterrupted by trees.

Standing on either side of the Prime Meridian, Greenwich. (Kat via Adobe)

MARITIME GREENWICH

In 1997, Maritime Greenwich was designated a World Heritage Site by the United Nations Educational Scientific and Cultural Organisation (UNESCO). Today, there are several places of scientific interest within the maritime and historical attractions.

OLD ROYAL NAVAL COLLEGE

London SE10 9NN

The Old Royal Naval College is a remarkable collection of baroque buildings and architecture that were mainly designed by Christopher Wren and Sir Nicholas Hawksmoor. The college was initially planned as a new palace for Charles II in 1660. However, Queen Mary II decided the site should be used for a hospital for seamen. Mary also decreed she should have a view of the river, which is why the Queen's House has an uninterrupted vista between the twin domes of the Royal Naval College. Unfortunately, she died of smallpox in 1694, but her husband William continued the project. The Royal Hospital for Seamen later became the Royal Naval College (1873–1998). Today, the complex is home to the Painted Hall, Chapel, and visitor centre.

The Old Royal Naval College is also a remarkable site rich in history. Beneath

The Old Royal Naval College Canaletto view.
(Old Royal Naval College Greenwich)

the buildings are the remains of Henry VIII's palace and his birthplace. In 1714, George I landed at Greenwich. It is also where Lord Nelson's body was brought after the Battle of Trafalgar in 1805 and where Queen Elizabeth II knighted Sir Francis Chichester after the 1967 circumnavigation of the world.

Visitor Centre

The visitor centre houses the ticket sales desk, shop, and Greenwich Tourism Office. The Pepys Building was erected between 1874 and 1883, and is where the Royal Naval College had tennis courts. There are exciting exhibitions in the building today, outlining the history behind the Royal Hospital for Seamen and the precision and architectural techniques used to build the Old Royal Naval College. Many techniques were applied from ancient Greek architectural styles.

Also housed in the Pepys Building is the Old Brewery, which was founded by Henry VIII. In later years it was used by the Royal Hospital for Seamen to produce the statutory three pints of ale for each pensioner. Today, the Old Brewery serves visitors a range of draught and bottled beers.

The Painted Hall

The Painted Hall is one of the highlights of any visit to Greenwich. Sir James Thornhill was commissioned to paint the ceiling of the Royal Naval College as a celebration of the protestant accession to the throne of King William III and Queen Mary II in 1688 and George I in 1714. Thornhill worked on the masterpiece from 1707 to 1726, and it is arguably his finest work. He was paid a pound for every square metre of the wall that was painted and £3 per square metre of the ceiling. As a result, over 3,700 sq m (40,000 sq ft) of walls and ceilings are covered in exquisitely detailed art.

The Painted Hall was initially intended to be a dining room for the residents of the Greenwich Naval Hospital. However, its glorious setting meant it soon became a tourist attraction and was used on special occasions. The artwork celebrates over 200 people set against a variety of backdrops. There are naval victories, scientists, and mythological and contemporary figures. When Thornhill painted his masterpiece, the Age of Enlightenment was at the forefront of thought. This was a time when experiments and scientific observation created new ways of looking at the world. Science was quickly seen as a critical element in the success of industry and commerce.

Astronomy was a particularly important science. Thornhill's artwork refers to Britain ruling the waves with its naval power. Mapping the stars was an essential skill of sailors so they could plot their position accurately at sea. Thornhill painted several great astronomers and scientists on

Painted Hall Ceiling, Old Royal Naval College, Greenwich. (Old Royal Naval College Greenwich)

Spirit of Architecture. Detail in Painted Hall. (Old Royal Naval College, Greenwich)

the ceiling, including his friend, John Flamsteed. Flamsteed was the first Astronomer Royal, based at the Royal Observatory in Greenwich, which was founded by Charles II in 1675. One of Flamsteed's most significant achievements was to plot the position of over 3,000 stars accurately. His star atlas, published posthumously, contained twenty-five maps of the constellations and was illustrated by Thornhill. Flamsteed is also shown predicting a solar eclipse on 22 April 1715, with his assistant Thomas Weston. The eclipse was highly significant at the time because totality, the complete coverage of the sun by the moon, passed over London. This event of totality was very rare and had not been seen since the time of King Alfred, and it would not be seen again until August 1999 in Cornwall.

The Painted Hall artwork also celebrates scientists who devised the theory that planets orbit the sun. In the painting, Copernicus holds what is known as his system. It demonstrates the sun at the centre of the solar system and shows Galileo holding a telescope. Although Sir Isaac Newton developed the universal laws of gravity and motion, proving that the Earth revolves around the sun, he is not depicted in the Painted Ceiling. It is possible Thornhill did not include him because Newton had fallen out with Flamsteed.

The Upper Hall was painted between 1718 and 1725 and shows Queen Anne with her husband, Prince George of Denmark. Prince George is pictured being handed the trident by Neptune, symbolising the supremacy of the Royal Navy. The family of George I is shown surrounded by allegorical bodies and a winged figure holds a scroll detailing the recent British defeat of the Spanish fleet at the Battle of Cape Passaro in 1718. In the bottom right-hand corner, James Thornhill is depicted holding out a hand, probably for payment.

Following his death at the Battle of Trafalgar in 1805, Admiral Lord Nelson's body lay in state in the Upper Hall, and a plaque marks the spot where the coffin was placed.

The Naval College Chapel

The Chapel of St Peter and St Paul was rebuilt in 1779 by James Stuart following a fire that destroyed Wren's original building. Naval themes

Chapel at the Old Royal Naval College. (Old Royal Naval College Greenwich)

dominate the chapel with its ornate ceiling and decorative art throughout the building, which was used by the Royal Hospital for Seamen. The rope designs and fouled anchor are symbolic of the Admiralty. The magnificent altarpiece depicting the shipwreck of St Paul on the Isle of Malta was painted by Benjamin West and is considered one of the finest eighteenth-century interiors in the world. Within the vestibule is a memorial to the Franklin Expedition in the 1840s involving the Erebus and the Terror. The memorial depicts hope before and despair following the event.

Underneath the Naval Chapel is a cellar housing the Victorian skittles alley. The room itself was once used by naval surgeons as an operating theatre. It was the ideal location in the days before anaesthesia was invented, as no one could hear the screams of the amputees in the underground room. There are deep grooves in the stone surrounding the windowsills. They served as knife-sharpening stones for the surgeons.

GREENWICH FOOT TUNNEL

www.visitgreenwich.org.uk/things-to-do/greenwich-foot-tunnel-p1373551

The Greenwich Foot Tunnel was built in 1902 and leads under the River Thames to Island Gardens on the Isle of Dogs, where there is an iconic view of Greenwich from across the water. The

Greenwich Foot Tunnel and Cutty Sark from the Thames. (Author)

tunnel was designed by Sir Alexander Binnie and replaced an unreliable ferry service. The tunnel took just three years to build, thanks to some exceptional engineering. The tunnel entrance is covered by a glass dome, and lifts or steps lead to the walkway.

QUEEN'S HOUSE

Romney Rd, London SE10 9NF

The Queen's House was designed by Inigo Jones in 1616 for Anne of Denmark (1574–1619), James I's queen. However, the building stopped in 1618, leaving the house one storey high. Queen Henrietta Maria, the wife of Charles I, resumed the project after 1629, when it was lavishly furnished up until the Civil War of 1642. The house was designed as a residential villa but has also served as the Greenwich Hospital School and as part of the National Maritime Museum. Today the palace is famous for its vast art collection, including a painting of Greenwich by Canaletto. From a scientific perspective, the remarkable Tulip Staircase was Britain's first geometric self-supporting spiral stair.

CUTTY SARK

King William Walk, Greenwich, SE10 9HT
www.rmg.co.uk/cutty-sark

One of the most famous ships in the world, this iconic tea clipper is now a museum in Greenwich. *Cutty Sark*

Cutty Sark. (Author)

is one of the first sights visitors get when they arrive in Greenwich from the river with her magnificent frame in a dry dock. The word 'clipper' came into use in the early nineteenth century and simply means a fast ship. Clippers were designed to be able to navigate ferocious seas off Cape Horn safely. However, tea clippers also needed to be fast in the water so they could unload cargo and obtain the best price for their goods. By 1861, an extra premium was offered for the first tea delivered to London, fuelling the need for faster ships. In 2007, a fire devastated *Cutty Sark*. However, the subsequent restoration has enabled better access to the ship.

Cutty Sark has several scientific features in her design, enabling her to move faster in the water. Her frame was made of wrought iron with planks of East India teak and American rock elm. The resultant small structure and thin hull enabled *Cutty Sark* to move at speed. In addition, by covering the hull in a brass alloy known as Muntz, the ship was protected from woodworm, barnacles and weed.

The hull on *Cutty Sark* is thinner than most other cargo ships. The narrow design enabled her to seamlessly cut through the water smoothly.

Cutty Sark is fitted with over 7km of rigging. Standing rigging supports the mast, whereas running rigging controls the sails and is made of hemp. When *Cutty Sark* was operational, she had a minimum of thirty-two sails comprising more than 3,000 sq m of canvas. The crew responded to the wind strength by adjusting the sails using the controlling ropes. This movement changed angles and enabled the ship to cross the wind or change course.

Visitors can explore *Cutty Sark* by visiting the museum and seeing the restored ship and its features. For those with a head for heights, a new experience climbing the rigging will get visitors up close to the ropes for a greater understanding of how sailors managed in the age of sail. Visitors can also enjoy a classic afternoon tea beneath the famous hull. Visits to *Cutty Sark*, including rigging and afternoon tea experiences, must be pre-booked.

NATIONAL MARITIME MUSEUM

Romney Road, London, SE10 9NF
www.rmg.co.uk

The National Maritime Museum is the most extensive collection of its kind in the world, and with Royal Museums, Greenwich holds more than 2.5 million items. There are many displays about nautical history within the museum. Scientific highlights include the following exhibits.

Scientific Instruments

On the ground floor, an exhibition of nautical instruments includes dividers

National Maritime Museum. (Author)

that measured distances on charts, surgical saws, and other implements used in battle on ships. In addition, some scientific instruments displayed at the National Maritime Museum were made in London. These include a pocket chronometer by John Arnold (1818–19), a theodolite made by Jonathan Sissons in 1737, and a polygonal telescope made in London in 1770.

Miss Britain III

Miss Britain III was the first boat to exceed 100mph on salt water and was designed by Hubert Scott Payne. She is covered in an alcad aluminium sheet, which reduces the drag and corrosion at sea. *Miss Britain III* was powered by a Napier Lion series VII B engine. She was built in 1933, and her construction inspired the use of Second World War motor torpedo boats. *Miss Britain III* is displayed in the Neptune Hall.

Tarbat Ness Light

In the Neptune Hall, the Tarbat Ness Optic is one of the prominent displays. It was designed by D.A. Stephenson in 1891. The light was built following the sinking of sixteen ships on the Moray Firth in 1826 and installed in 1892. The Optic remained with the lighthouse until 1985. It consists of lenses, prisms, and reflections that form a single powerful beam. Each lighthouse has its unique signal. Tarbat Ness beamed four times every thirty seconds and could be seen 15 miles away.

Caird Library

The Caird Library is located within the National Maritime Museum, and visits must be pre-booked, including library membership. The library is named after Sir James Caird, who sponsored Franklin's expedition to the Arctic. It contains over 100,000 books, 20,000 periodicals, including 200 current titles, 20,000 pamphlets, and many rare books. In addition, there are ships' log books, records on horology and navigation, and astronomy. The Caird Library is also home to over 80,000 maps, some dating to the fifteenth century. In addition, the museum has records from the mutiny on the *Bounty*, guides to nineteenth-century naval surgery and prominent historical maritime events.

SCIENCE ON THE STREET

AVERY HILL PARK

Bexley Rd, London SE9 2PQ

Avery Hill Park is an open parkland area and home to the magnificent winter garden owned by Greenwich University. In 1888, Colonel John Thomas North bought the parkland and developed the Italian Gardens, the Winter Garden, and much of the park area visible today. He was known as the 'Nitrate King' because he imported nitrates from Chile, which were used in industry and agriculture. In particular, he established a monopoly in saltpetre used in fertilisers. North died suddenly in 1896. The Royal Borough of Greenwich now owns the park.

ENDERBY WHARF

Olympian Way, London, SE10 0TA

Enderby Wharf on the Greenwich Peninsula is associated with the Atlantic cables used for telecommunications between Europe and other parts of the world. Today the site is a housing estate, but there are reminders of its historical past. The technology was responsible for transforming communications and how the City of London and finance sectors operate. The first attempt at laying a submarine cable was in 1857. Today, there is little left of the cable industry that was once here apart from grey steel towers and wheel engine cable as a reminder of what was once a pioneering development in submarine cabling. There is also a public artwork called Lay Lines, depicting the laying of submarine cables.

SCIENTIFIC MEMORIALS AND STATUES

ST MARGARET'S CHURCH

Lee Terrace, London, SE13 5DL
www.stmargaretslee.org.uk

Although St Margaret's Church is not in the parish of Greenwich, it has some scientific heritage. And there has been a church on site for over 900 years. The medieval church ruins are on the north side of Lee Terrace. Close by is the tomb of Edward Halley (1656–1742), the second Astronomer Royal, who calculated the orbit of the comet named after him. Two other Astronomers Royal are buried nearby: Nathaniel Bliss and John Pond.

WALTER AND ANNIE MAUNDER BLUE PLAQUE

69 Tyrwhitt Road, Brockley, SE4 1QE

Walter and Annie Maunder were two prominent astronomers. They are best known for their work in eclipses and sunspots, promoting amateur astronomers and women's work in astronomy. Both Walter (1851–1928) and Annie Maunder (1868–1947) worked at the Royal Observatory in Greenwich; however, Annie was unable to work as a professional astronomer at the time because the profession was not open to women. They lived at 69 Tyrwhitt Road, Brockley, from 1907 to 1911. While living there, they wrote *The Heavens and Their Story* (1908), which captured their stargazing at nearby Hilly Fields and promoted amateur astronomy. Both Maunders have three craters on the moon named after them.

7

KENSINGTON, CHELSEA, AND WEST LONDON

Dominated by some of the most famous museums in Britain, Kensington and Chelsea is full of scientific attractions. From the famous blue whale at the Natural History Museum to the Chelsea Physic Garden, this part of London has many scientific-themed places to go.

FOCUS ON THE SCIENCE MUSEUM

The Science Museum, Exhibition Road, South Kensington, London, SW7 2DD
www.sciencemuseum.org.uk

The Science Museum is one of the most famous museums in London and is a major visitor attraction. It was founded in 1857 and holds one of the world's most prominent collections of scientific items. The museum is free to enter; however, some special exhibitions attract an additional charge, and donations are welcome.

Tips for Visiting

Although the museum is free, it is best to pre-book a time slot for entry, particularly during the school holidays. School parties visit the museum frequently, so going earlier or later in the day helps avoid those peak times when the building can get noisy.

The Science Museum is very accessible, with lifts and interactive displays. There are also dedicated visiting times for people with special sensory needs, which are advertised on the website.

There are five floors of scientific displays, so trying to see everything in a day is overwhelming. Instead, focus on a couple of areas or head for the highlights. If you plan to see a special exhibition within the Science Museum, you will usually need to pre-book a time slot in addition to the main entry timing.

It is easy to get lost in the Science Museum, so pick up a map at the entrance so you can see the various departments.

Black Arrow at the Science Museum. (Science Museum Group)

Ground-Floor Highlights

The Ground Floor is the Energy Hall, where several steam engines are on display highlighting the impact of the Industrial Revolution. The exhibits include Old Bess, the oldest surviving James Watt Beam Engine, and items from Watt's garret workshop. The innovative gallery Space Exploration is also located on the ground floor. Highlights include Tim Peake's spacecraft, a genuine moon rock, and satellites. Another exhibit are items from Professor Stephen Hawking's office. The Making the Modern World Gallery includes Charles Babbage's Difference Engine No. 1 and the first Apple computer. Within Exploring Space is a replica of *Eagle*. In 1969, this lander took astronauts Armstrong and Aldrin to the moon. Other features include space rockets suspended from the ceiling and a model of the Beagle 2 Mars lander.

The Imax Ronson Theatre and Hans Rausing Lecture Theatre are also on the ground floor.

First-Floor Highlights

The Wellcome Galleries fill most of the first-floor area and highlight medical innovations across the centuries. Interactive displays outline how dissection and the work of John Hunter informed the development of medicine and surgery. Exhibits also show how disease is prevented. In addition, there are iconic items such as Laennec's stethoscope, Joseph Lister's watch, and Rosalind Franklin's Photo 51 of a DNA sequence. Other displays outline phrenology, the development of electrocardiograms, and quarantine.

The David Sainsbury Gallery on level one has an interactive display of career opportunities in science for technicians.

Second-Floor Highlights

Level two is home to the Winton Gallery, which was designed by Dame Zaha Hadid, and specialises in mathematics. The design was inspired by the equations used to calculate airflow in the aircraft industry and has a Handley Page aircraft at the heart of the gallery. This exhibition has been constructed to demonstrate how mathematics integrates everyday life, from gambling

Entrance to the technicians exhibition at the David Sainsbury Gallery. (© Science Museum Group)

to the economy and predicting tides. Exhibits include medical statistics such as Florence Nightingale's cockscomb diagram, a display of how the economy works, and early computing from pioneers like Ada Lovelace. In addition, volunteer guides give impromptu talks in the Winton Gallery.

The Information Age Gallery has exhibits dedicated to telling the story of how technology and communication have changed over 200 years.

Third-Floor Highlights

Level three has a large section dedicated to how science transformed London. Science City includes the work of Isaac Newton and Robert Hooke, including how the Monument was built. Newton's reflecting telescope is on display. There is also King George III's Philosophical Table, where experiments in force and motion were demonstrated. A section of level three is dedicated to clockmakers in London and is the oldest collection of its kind in the world.

Another section of level three is dedicated to aviation, known as Fly Zone. This exciting collection includes Amy Johnson's Gipsy Moth, a slice of a Boeing 747, and a Supermarine S6B racing seaplane. Overhead walkways enable visitors to get up close to the aircraft.

LONDON LANDMARK

THE NATURAL HISTORY MUSEUM

Cromwell Road, London, SW7 9BD
www.nhm.ac.uk

The Natural History Museum opened to the public in 1881, but it originated from the collections of Sir Hans Sloane in 1753. Sloane was a renowned physician who treated the wealthy, which enabled him to travel and collect specimens from across the world. After he died in 1753, the terms of his will enabled Parliament to buy the massive collection of over 71,000 items for £20,000, which was significantly undervalued. The Government then built the British Museum to house the collection.

In 1856, Sir Richard Owen became curator of the natural history collection. He would become famous for coining the name for dinosaurs, but he also influenced the British Museum governors to build a dedicated museum for natural history.

Alfred Waterhouse designed the building on Cromwell Road, using terracotta to construct it as it was more resistant to the London climate in Victorian times. The building is an exquisite example of Romanesque architecture, but its structure is the

Blue Whale Hintze Hall at the Natural History Museum. (Trustees of NHM)

Hintze Hall, Natural History Museum. (Trustees of NHM)

legacy of Owen's ideas that museums should be freely accessible and able to house large specimens. One of the most enchanting features of the Natural History Museum is the distinctive animal and plant carvings and reliefs all around the building. They were designed by Alfred Waterhouse, who checked the accuracy of the features with scientists. In addition, the gallery roof is covered with painted tiles depicting plants from around the world.

In 1986, the museum incorporated the Geological Museum of the British Geological Survey and over 30,000 specimens. The Darwin Centre opened in 2009. It houses the museum collection and working scientists. The Cocoon structure displays the museum's most important specimens, and visitors can watch scientists at work.

Museum Highlights

The museum is split into colour-coded zones.

Blue Zone

In the Blue Zone, the Dinosaur Gallery is very popular. Pre-booked timed tickets to see Dippy the Diplodocus are advisable, especially at weekends. Other features of the Blue Zone include the world's largest mammals, displays of fish, amphibians, and marine invertebrates, and a quiet zone.

Green Zone

The iconic blue whale is the centrepiece of the Hintze Hall in the Green Zone, where visitors can walk underneath the huge skeleton or view from above. There are oak cabinets full of minerals and a prolific fossil collection. Among the bird exhibits is a dodo. The Vault contains some of the museum's more unusual and unique exhibits. They include a piece of the Winchcombe Meteorite that fell to earth in 2021 and a sample of Kernowite, first described by scientists in 2020.

Red Zone

The Red Zone has exhibits about volcanoes and eruptions, and a Stegosaurus fossil skeleton in the Earth Hall. The Earth's Treasury Gallery has displays of rocks and minerals. There are also exhibits on human evolution in the Red Zone.

Orange Zone

The Cocoon structure is located within the Orange Zone. Another feature is the Zoology spirit building, where 23 million specimens are stored. Behind-the-scenes tours are available but must be pre-booked.

Other events at the Natural History Museum include special exhibitions, sleepovers, and volunteering.

The giraffe at the Natural History Museum.
(Trustees of NHM)

BOTANICAL AND NATURAL HISTORY HERITAGE

CHELSEA PHYSIC GARDEN

66 Royal Hospital Road, Chelsea, London, SW3 4HS
www.chelseaphysicgarden.co.uk

The Chelsea Physic Garden was established by apothecaries in 1673 and still occupies the 4 acres of land on the River Thames purchased by Dr Hans Sloane so the work could continue. It was developed by the Worshipful Company of Apothecaries to grow plants for medicinal use and is among the oldest botanical gardens in Britain. Some 350 years after the apothecaries began sowing seeds, the garden is stocked with over 5,000 medicinal, edible, herbal, and valuable plants.

Visitors can walk around the gardens or enjoy a guided tour. In addition, workshops on subjects such as writing, gardening and health are held.

MUSEUMS WITH SCIENTIFIC, MEDICAL AND ENGINEERING HERITAGE

THE ALEXANDER FLEMING LABORATORY MUSEUM

St Mary's Hospital, Praed Street, London, W2 1NY
www.imperial.nhs.uk/about-us/who-we-are/fleming-museum

The Alexander Fleming Laboratory Museum is located within the grounds of St Mary's Hospital and in the exact place where this eminent scientist discovered penicillin in 1928. A reconstruction of his laboratory gives an insight into how he worked and the remarkable story of the chance discovery of a lifesaving drug. Fleming was an observational scientist, leaving Petri dishes in his laboratory for weeks to see what would happen. After returning from a holiday in September 1928, he noticed one of his dishes had grown a fungus but something was clearing the bacteria. He had discovered penicillin.

Visits to the museum must be pre-booked. Access is limited to the stairs due to the historic nature of the building.

Alexander Fleming Museum. (Author)

ROYAL HOSPITAL CHELSEA

Royal Hospital Road, SW3 4SR
www.chelsea-pensioners.co.uk

The Royal Hospital was founded in 1682 by Charles II as a home for retired or disabled soldiers. It was designed and built by Sir Christopher Wren, with the Great Hall a masterpiece in baroque architecture. Today, the Royal Hospital is home to the iconic Chelsea Pensioners, where the veterans receive care and social support. There are guided tours led by a Chelsea Pensioner, which tell the story of this renowned institution and give an insight into their personal experience. All tours must be pre-booked.

VICTORIA AND ALBERT MUSEUM

Cromwell Rd, London, SW7 2RL
www.vam.ac.uk

The Victoria and Albert Museum primarily comprises applied arts, design, and decorative artwork. Some items are connected with science including a sculpture of Albert Einstein in the entrance area.

A set of notebooks containing scientific and mathematical diagrams by Leonardo da Vinci, known as Codex Forster 1, are displayed in the museum and show his hydraulic designs.

Mechanical clocks are also displayed at the V&A, including a mechanical globe clock from 1580. Astrolabes, sundials, an astronomical compendium, and other instruments also feature in the collection.

The Townsend Collection of gems shows a display of natural stones used in modern jewellery.

THE DESIGN MUSEUM

224-238 Kensington High Street, London, W8 6AG
https://designmuseum.org

When the Commonwealth Institute closed its headquarters in 2002, its Kensington High Street building underwent a dramatic transformation and is now the Design Museum. Today, it is one of the world's leading museums on contemporary design and has both permanent and temporary exhibitions of innovative work. Of particular note is the distinctive roof design in the shape of a parabola. The double paraboloid curve was intended to portray optimism and progress and was based on the designs of the Spanish engineer Felix Candela. Many of the techniques used in the roof building had not been attempted in Britain before. Originally, poured concrete was going to be used for the roof; however, it was shown that this needed to be stronger. So instead, pre-cast concrete beams were used, which created the radiating pattern on the roof.

Exhibits include a range of designs, from robots to household items and applied arts.

RESTAURANTS, PUBS AND BARS CONNECTED TO SCIENCE

SCIENCE-THEMED AFTERNOON TEA AT THE AMPERSAND HOTEL

10 Harrington Road, London, SW7 3ER
https://ampersandhotel.com

Afternoon tea is a treat, but the Ampersand Hotel injects a little fun with its science-themed teas. Choose from two scientific-themed afternoon teas, which will keep curious children (and adults) entertained as they eat. The science afternoon tea features jam in a petri dish, an experiment with lemonade, planet-shaped cakes, and more. Alternatively, the Jurassic-themed

Afternoon tea at the Ampersand Hotel. (Author)

tea includes a volcanic eruption and a dinosaur egg basket.

The Ampersand Hotel also has nature-themed rooms aligned with the themes of the nearby museums.

COLONNADE HOTEL

2 Warrington Crescent, London, W9 1ER
www.colonnadehotel.co.uk

The Colonnade Hotel is opposite Warwick Road Underground Station. It has historical significance as it is the birthplace of Alan Turing. The present-day hotel started out as two private residences before being converted to a maternity hospital in 1886. Turing was born on 23 June 1912 and became a successful mathematician and cryptanalyst. By 1935, the building was a hotel and in 1938 hosted Sigmund Freud, his wife, and daughter Anna. The family stayed here as guests while their Hampstead home was being refurbished.

Alan Turing blue plaque at the Colonnade Hotel. (Author)

FOUNTAINS ABBEY PUB

109 Praed Street, Paddington, Greater London, W2 1RL
www.greeneking-pubs.co.uk/pubs/greater-london/fountains-abbey

The Fountains Abbey is opposite St Mary's Hospital on Praed Street. It was frequented by Alexander Fleming and has memorabilia relating to the scientist on the walls.

LANDMARK HOTEL

222 Marylebone Road, London, NW1 6JQ
www.landmarklondon.co.uk

The Landmark Hotel in Marylebone is one of the grandes dames of the capital's hotels. It has seen several historic events in its history. From a scientific perspective, when Alexander Fleming discovered penicillin, a charity event was held at the hotel to raise funds for St Mary's Hospital.

THE BLETCHLEY

Chelsea Funhouse, 459 King's Road, Chelsea, London
www.thebletchley.co.uk/home

The Bletchley is a cocktail bar with a difference. Using replica Second World War Enigma machines and Sherlock Holmes-style protocols, guests get involved in code-breaking to create a personalised cocktail at the bar. Naturally, the recipes stay a secret, just as all codes should. However, there is a standard menu for people who simply want a drink.

ENGINEERING EXPERTISE

Paddington Basin has some interesting bridges with physics and engineering features.

FAN BRIDGE

www.thisispaddington.com/article/fan-bridge-revealed

The Fan Bridge crosses the Grand Union Canal and is a remarkable piece of engineering. It was designed by Knight Architects and completed in 2014. A cantilevered deck is hinged at the northern end of the bridge. It is raised using hydraulic jacks, and the action is similar to a Japanese fan opening, giving the bridge its name.

The steel beam that forms the deck is split into five fingers, each with bearings that make it open in sequence, with the first rising to an angle of 67 degrees. The last finger gives the required clearance over the canal. Counterweights help reduce the energy required to raise the bridge structure. The kinetic sculpture has an eye-catching silhouette as it forms into the fan shape. Bridge openings occur at specific times of the week, listed on the This is Paddington website.

Fan Bridge, Paddington. (Author)

ROLLING BRIDGE

www.thisispaddington.com/article/heatherwicks-rolling-bridge

The Rolling Bridge was designed by Heatherwick Studio and was completed in 2004. Designers wanted a bridge that opens in a special way. The deck of the bridge is constructed in eight triangular sections, which have seven pairs of hydraulic rams within the balustrades, each corresponding to the joints on the deck. When the rams extend, they push up the handrail, which makes the bridge curl up.

One underground master ram powers the fourteen rams. Once the curling motion is complete, the two bridge ends touch, forming an octagon.

Rolling Bridge, Paddington. (Author)

LEARNING OPPORTUNITIES

ROYAL GEOGRAPHICAL SOCIETY

1 Kensington Gore, London, SW7 2AR
www.rgs.org

The Royal Geographical Society has a prolific collection of resources for schools and professionals, aimed at progressing the study of geography. The RGS also has events and lectures on aspects of geography and exploration throughout the year.

SCIENTIFIC MEMORIALS AND STATUES

Kensington and Chelsea have several statues and memorials relating to science.

KENSINGTON GARDENS

Albert Memorial Road, London
www.royalparks.org.uk/parks/kensington-gardens

The area in and around Kensington Gardens has several scientific connections.

Albert Memorial

The Albert Memorial was designed by George Gilbert Scott and unveiled in 1872 to commemorate the death of Prince Albert, Queen Victoria's husband. Albert is depicted holding a copy of the Great Exhibition catalogue, which he helped organise. The memorial celebrates Victorian achievements. At the four corners of the memorial, there are sculptures of personifications representing Victorian industrial art and science (engineering, commerce, agriculture, and manufacturing). The canopy has an ornate design with mosaics and sculptures. The pillars and niches of the canopy feature eight statues representing the practical arts and sciences: Geometry, Astronomy, Geology, and Chemistry (on the four pillars) and Medicine, Philosophy, Physiology, and Rhetoric (on the niches).

Physical Energy Statue

The bronze statue of a man on horseback is known as Physical Energy and was designed by George Frederic Watts (1817–1904). The artist saw Physical Energy as an allegory for the human need to look constantly for new challenges and to look to the future. The cast in Kensington Gardens is also a memorial to Watts and was erected by his wife in 1907.

Jenner Statue, Kensington Gardens. (Author)

Jenner Memorial

At the Bayswater Road end of Kensington Gardens is the delightful Italian Garden. Edward Jenner (1749–1823) was a doctor from Berkeley, Gloucestershire, who invented the smallpox vaccine. The statue was erected in 1858 in recognition of his pioneering work and was designed by William Calder Marshall.

JAMES CLERK MAXWELL BLUE PLAQUE

16 Palace Gardens Terrace, Kensington, London, W8 4RP

James Clerk Maxwell (1831–79) was a physicist responsible for the theory of electromagnetic radiation. He brought several theories on electricity, magnetism and light together to form the concept of electromagnetic theory. A blue plaque has been placed on the wall of 16 Palace Gardens where he once lived.

STATUE OF DR HANS SLOANE, DUKE OF YORK SQUARE

Duke of York Square, London SW3 4RB.

Dr Hans Sloane, the physician to the wealthy, amassed a massive collection of specimens and artefacts worldwide. His collecting helped found the British Museum, the Natural History Museum, and the Chelsea Physic Garden.

LORD KELVIN BLUE PLAQUE

15 Eaton Place, Belgravia, London, SW1X 8BN

Lord Kelvin (1824–1907) was a physicist who invented the Kelvin Scale of temperatures designed to measure low temperatures. He also installed a telegraph cable under the Atlantic Ocean in 1866. Lord Kelvin also invented the magnetic compass and binnacle, further improving it with two compensating spheres known as Kelvin's balls. He lived at 15 Eaton Place.

ROSALIND FRANKLIN BLUE PLAQUE

Donovan Court, 107 Drayton Gardens, Chelsea, London, SW10 9QS

Rosalind Franklin was a British scientist who pioneered the study of molecular structures. She is most famously known for her research into DNA molecules at King's College, which was the catalyst in helping Watson and Crick identify the structure of DNA in 1953.

STATUES OF ALAN TURING AND MARY SEACOLE

St Mary's Terrace, London, W2 1SY

At the southern end of St Mary's Terrace, near the bike racks, are two steel statues of Alan Turing and Mary Seacole. Turing is known as the father of computer science and was born in the area. Mary Seacole nursed wounded soldiers during the Crimean War.

STAINED GLASS MEMORIAL TO ALEXANDER FLEMING

St James's Church, Sussex Gardens, Paddington, London, W2 3UD
https://stjamespaddington.org.uk

St James's Church in Paddington has a beautiful stained glass window dedicated to the work of Alexander Fleming. It was installed in 1952.

BROMPTON CEMETERY

Fulham Road, London SW10 9UG

Over 200,000 people lie in rest at Brompton Cemetery. Among the graves are several associated with

Alan Turing and Mary Seacole statues near Paddington. (Author)

science and medicine. These include Dr John Snow, famed for his work on epidemiology. Sir William Crookes is also buried at Brompton Cemetery. He was a renowned chemist who discovered thallium and produced polarising lenses for sunglasses, among other developments. Dr Benjamin Golding founded Charing Cross Hospital. John Peake Knight was a pioneer in railway engineering safety.

SOUTH-WEST LONDON

South-West London is most famous for the Royal Botanic Gardens at Kew, where plant hunters founded the extraordinary horticultural collection and where scientists today look for ways to protect the botanical world. However, the area also has some of the finest engineering examples in London. There are Cornish engines at the London Museum of Water and Steam, restored engineering at Kempton Park, and the redesign of Battersea Power Station has created an iconic new venue in London.

LONDON LANDMARK

ROYAL BOTANIC GARDENS KEW

Kew, Richmond, London, TW9 3AE
www.kew.org

Nature and the seasons bring change to Kew Gardens, so there is always something new to see. Whether the vibrant autumnal colours bring joy to a walk through the gardens or an exhibition highlights new work by scientists to protect the environment, Kew is constantly refreshing to visit.

Tips for Visiting

The Kew Gardens website details events, exhibitions, and planned closures. It also details which sections of the gardens look particularly good at the moment so you can plan your visit. Online pre-booked tickets cost significantly less than a walk-up ticket. There are several entrances to the gardens, with the main one at Victoria Gate. If you plan to take a break from the gardens and explore other attractions in the area, you can re-enter Kew the same day by showing your ticket at any of the entry points. Some areas of Kew Gardens close during winter or for special events, so check the website before travelling.

History of Kew Gardens

Kew Gardens was founded in 1759, just as Britain began learning more about

Palm House, Kew Gardens. (Visit London/Jon Reid)

the world through science. The gardens initially housed exotic plants brought back to Britain by explorers. When Sir Joseph Banks joined Captain Cook's famous voyage on HMS *Endeavour* from 1768 to 1771, he was inspired by the plant life he saw. On his return, he became Kew's first (unofficial) director, with the ambition that it should become the best botanic garden of its time. Over the centuries, Kew has been enriched by the plant hunters, bringing exotic plants to the botanical gardens and expanding its extraordinary collections.

However, Kew's purpose evolved as Britain developed colonies and needed plants of economic value. In 1840, Kew Gardens was given to the nation by the Crown and opened to the public. People came in their thousands. Sir William Hooker became the first official director and further developed the gardens. Both the Palm House and Temperate House were added to Kew Gardens at this stage and are iconic examples of Victorian architecture. The public could now admire tropical plants in the famous glasshouses. Sir Joseph Dalton Hooker succeeded his father as director and, as one of the most prolific plant hunters, made Kew the centre of a global exchange centre for plants. Seeds for rubber trees received in Kew in 1876 were planted so seedlings could be sent to Malaysia and Sri Lanka to generate their industries.

Today, the Royal Botanic Gardens at Kew offer training to gardeners as it has done since 1859. It is also at the centre of scientific efforts to protect plants, especially endangered species and threats from environmental change.

Highlights of Kew Gardens

There is much to see and do within the Royal Botanic Gardens at Kew. You may need more than one visit to see it all, but it is naturally somewhere to return to for viewing and experiencing alternative perspectives throughout the year.

Palm House

Stepping into the iconic Palm House feels and smells like a journey into a tropical rainforest. Walk around the glasshouse in the steamy heat where there are endangered plants and those valuable to society. One of the plants flourishing in the glasshouse is the palm-like Cycad (Encephalartos altensteinii) and the oldest pot plant in the world. It was brought to Britain from South Africa in 1775 by Francis Masson. The plants are used by scientists at Kew to try to develop more sustainable harvests and other research.

Waterlily House

Giant waterlilies are famous at Kew. The remarkable giant waterlily Victoria Amazonica is grown from seed annually, and the leaves can reach up to 2.8m in dimension. The plants are hand pollinated at Kew and are a joy to see.

Temperate House

The Temperate House is the world's largest glasshouse and home to over 1,500 endangered species. All the plants need to be maintained at 10°C to thrive. The work in the Temperate House is helping scientists find solutions to rescuing endangered plant species.

Japanese Garden

The delightful Japanese Garden is at its finest in autumn with the rich colours of the leaves and in springtime with blossoms. The gardens incorporate a Peace Garden and a Garden of Harmony.

Treetop Walkway

A canopy walkway 18m above ground allows visitors to admire the many trees from a height. From various vantage points, visitors can see the complex ecosystems of trees, including lichens and birds.

The Hive

The Hive is a remarkable piece of contemporary art designed to simulate being in a beehive. LED lights glow, mimicking the vibrations bees make, and even the music plays in the key that bees make when buzzing.

The Great Broad Walk Borders

Originally designed as an elegant promenade to the Palm House, the Great Broad Walk Borders are 320m in length and have a rich variety of plants at all times of the year. They are the longest herbaceous borders in Britain, and the rainbow of colour is designed to inspire gardens and gardeners.

The Great Pagoda

Another iconic sight in Kew Gardens is the Great Pagoda. It was completed in 1762 for Princess Augusta, who founded the gardens. The pagoda was designed as a folly where visitors could admire views of London, and still do to this day.

Edible Science: Kew's Kitchen Garden

Edible plants and heritage varieties of vegetables are grown in the Kitchen Garden. Scientists use the area to look at how vegetables can be produced sustainably and more resistant to climate change.

The pagoda at Kew Gardens. (Visit London/ Jon Reid)

ENGINEERING EXPERTISE

BATTERSEA POWER STATION

Circus Rd W, Nine Elms, London, SW11 8AL
https://batterseapowerstation.co.uk

The iconic Grade II listed Battersea Power Station has been an integral part of the London landscape for years. It was designed by Sir George Gilbert Scott, and construction of the art deco building began in 1929. At its peak, the coal-fired power station produced a fifth of London's electricity. Battersea closed in 1983, however, a recent innovative plan to regenerate the area has brought new life to the area with a shopping mall, exhibits, a cocktail bar, and a chimney lift. Much of the original architecture has

Battersea Power Station. (The other key on Pixabay)

been integrated into the new design, giving Battersea Power Station a unique and striking atmosphere.

Highlights of Battersea Power Station

There are many shops, restaurants and a cinema in the new Battersea Power Station. Other highlights include the following features.

Power of Place Exhibition
This exhibition, located in the North Atrium, Ground Floor, outlines the story of Battersea Power Station and how it has transformed from the closure to a new cultural centre.

Heritage and Learning Hub
Visitors can see a historical model of the Battersea Power Station at the Grosvenor Railway Arch and understand how it powered London.

Lift 109
https://lift109.co.uk

One of the unique experiences at Battersea Power Station is the opportunity to take a lift up the north-west chimney for a panoramic view of London. The experience starts in the iconic Turbine Hall, where an exhibition and multimedia displays outline the rich heritage of the building. VIsitors then ascend 109m through the chimney to views across London.

Control Room B Cocktail Bar
www.controlroomb.com

Control Room B is a unique setting for a cocktail bar with original dials and other features. The turbine-inspired radial structure is a nod to the heritage of Battersea Power Station. Guests can sip a cocktail and be transported back to the 1950s in this inspirational bar.

LONDON MUSEUM OF WATER AND STEAM

Green Dragon Lane, Brentford, London, TW8 0EN
https://waterandsteam.org.uk

The London Museum of Water and Steam is on the site of the old Kew Bridge Waterworks, which opened in 1838. By 1900, Kew Bridge was a significant part of maintaining London's water supply. However, by 1944, maintaining a system with steam engines had become prohibitively expensive, so Kew Bridge was designated a museum site to preserve the engineering heritage. The Kew Bridge Engines Trust was formed in 1974 and continued to preserve the engines, and was rebranded to the London Museum of Water & Steam in 2014.

At the heart of the museum is its magnificent collection of steam-pumping engines. The Cornish engines are within their original engine houses. There are also rotative engines collected from around the country by the museum trust.

The Reader electricity generating engine at the London Museum of Water and Steam. (Author with permission from the London Museum of Water and Steam)

Cornish Engines

The three engines in the museum collection were made in Cornwall, but Cornish refers to the engine's operating cycle. In a Cornish engine, pumping is done using a falling weight lifted by the engine. The weight is above the pump, which is connected to a beam. A piston is at the opposite end of the beam. A combination of steam pressure above and a vacuum below the piston lifts the weight. As the weight falls during the pumping stroke, the piston returns to the top of the cylinder due to an equilibrium valve opening to allow steam to pass.

The London Museum of Water and Steam has the most extensive collection of Cornish engines in the world.

The Rotative Engines

The Steam Hall has four large rotative engines and several ancillary ones. The crank converts the linear motion of the piston into rotatory motion in the rotative engine. Finally, both cranks and flywheels balance the loads and thrusts so the engines can operate smoothly.

The Splashzone children's play area at the London Museum of Water and Steam. (Author with permission from the London Museum of Water and Steam)

The museum also has items relating to the water supply. Outside, there is an area for children where the play items have been designed with engineering features.

KEMPTON STEAM MUSEUM

Kempton Park Waterworks, Snakey Lane, Hanworth, Middlesex, TW13 6XH
https://kemptonsteam.org

Kempton Park Pumping Station has supplied fresh water to London since 1906. Today the site is managed by Thames Water PLC, but the site is much more. It is also home to the Kempton Park Great Engines Trust, who have preserved the triple-expansion steam engines and turbines that served London for more than fifty years. Today, the steam engines are open to the public on defined steam up and heritage days.

The triple-expansion engines weigh 800 tons, stand 62ft tall, and are known as the Sir William Prescott and Lady Bessie Prescott engines. Their top speed is 25.4 revolutions a minute. Kempton Park is also home to a Mercury Arc Rectifier, which converts the incoming 415-volt alternating current from the National Grid into 200 volts of direct current for compressors and engines in the museum. The rectifier originally came from the Royal Opera House and has been restored. Kempton Park also has various instruments, including venturi mercury readers and pressure gauges.

MUSEUMS WITH SCIENTIFIC, MEDICAL AND ENGINEERING HERITAGE

THE MUSICAL MUSEUM

399 High Street, Brentford, TW8 0DU
www.musicalmuseum.co.uk

At one time, self-playing musical instruments were popular in private homes and public places of entertainment. In 1963, Frank Walter Holland, an electronic engineer by trade, noticed that many automatic musical instruments were being discarded and abandoned, so he set about collecting as many as possible. His foresight led to the creation of an eclectic collection in a local church until it was moved to its current location near Kew Bridge. Today, the Musical Museum contains one of the world's finest collections of automatic musical instruments, including exquisite music boxes, barrel organs, pianolas, and orchestrions (a machine that plays music that sounds like an orchestra).

The Musical Museum at Brentford. (Author)

Holland's idea was to create a place where people could both hear and see the instruments on display. Today, visitors can either walk around the museum independently or take a guided tour where many of the organs and music boxes are demonstrated.

Some of the highlights of the Musical Museum include the collection of music boxes, some dating from the eighteenth century. The music boxes operate with a cone that has a set of metal levers and a cylinder with metal spikes. As the spikes hit the teeth, they produce a sound. The teeth are of different lengths; the longer the length, the deeper the sound.

Another unusual item in the museum is the eerie-sounding theremin. This remarkable instrument is played by hand but without being touched. The theremin produces a single note that is changed by the player moving their right hand up and down in front of an antenna. The volume is controlled by the player's left hand moving over a loop antenna. Leon Theremin invented the instrument in 1920 in Russia and marketed his work in Europe and America. However, on his return to Russia in 1938, he disappeared under the Stalin regime and was sent to a labour camp where he was used to develop Soviet bugging devices. He returned to the USA in 1991 and died in Moscow in 1993 aged 97.

The museum's highlight is the Mighty Wurlitzer organ dating from 1928, which was used in cinemas and accompanied silent movies.

THE SCIENCE OF DISTILLING

SIPSMITH GIN TASTING TOURS

83 Cranbrook Road, Chiswick, London, W4 2LJ
https://sipsmith.com

Gin is made using the science of distillation. Chiswick is home to Sipsmith, who use pot stills (Prudence, Constance, and Verity). The still is connected to a condenser, which is also connected to a barrel to collect the distillate. Next, botanicals and the base spirit are added to the still's body. Water does not boil until it reaches 100 degrees, whereas alcohol starts to boil at 73 degrees. This means that the alcohol in the spirit will become a gas and rise while the water stays in the still's bottom. As a gas, the alcohol rises to the top of the still and through the condenser (filled with cold water), where it cools. The final distillate trickles into the barrel, where it can either be retained or returned to the still for a second and higher concentration of distilling.

Sipsmith offer guided tours of their distillery to see how gin is produced. Tasting is included. Tours are for over 18s only and must be pre-booked.

BOTANICAL AND NATURAL HISTORY HERITAGE

LONDON WETLANDS CENTRE

WWT London, Queen Elizabeth Walk, Barnes, London, SW13 9WT
www.wwt.org.uk/wetland-centres/london

Wetlands are critical in maintaining biodiversity and protecting the environment from the threat of climate change. Up to 40 per cent of the world's animals and plants depend on wetlands. The London Wetlands Centre has a huge variety of species, including kingfishers, wading birds, sand martins, lizards, and water voles. There are walks, observation points, and talks at this oasis on the edge of central London.

SCIENTIFIC MEMORIALS AND STATUES

SIR WILLIAM HOOKER AND SIR JOSEPH HOOKER BLUE PLAQUE

49 Kew Green, Kew, London, TW9 3AA

A blue plaque marks the house where Sir William Hooker (1785–1865) and Sir Joseph Hooker (1817–1911) lived. They were pioneers of plant hunting and in developing the Royal Botanic Gardens at Kew.

ABDUS SALAM BLUE PLAQUE

8 Campion Road, Putney, London, SW15 6NW

Dr Abdus Salam (1926–96) was one of the most influential physicists of the twentieth century. He was born in Pakistan and is a Nobel Prize winner and the first Muslim to win one in the sciences. Dr Salam developed the theory demonstrating how electromagnetic and weak forces may be considered manifestations of a single, more fundamental force. The concept paved the way for the discovery of the Higgs Boson particle. Abdus Salam was also an ambassador for promoting physics in developing countries. He lived at the house in Putney for forty years.

RESTAURANTS, HOTELS AND PUBS WITH A SCIENCE CONNECTION

ROOM 2 HOMETEL, CHISWICK

10 Windmill Rd, Chiswick, London, W4 1SD
https://room2.com/chiswick

Room 2 Hometel was inspired by the heritage art deco architecture in Chiswick. However, it is also the first hotel in the world to be 100 per cent carbon neutral. As a result, the hometel uses over 80 per cent less energy than a conventional hotel and is plastic free. Bees even make honey on the roof.

NORTH LONDON

North London is characterised by distinct communities, from the suburban feel to Heath Robinson's Pinner to healthcare teams working to reduce health inequalities in Islington and Camden. There are engineering landmarks, too, interspersed with diverse communities that make up London's vibrant culture in North London.

LONDON LANDMARK

HIGHGATE CEMETERY

Swain's Ln, London N6 6PJ

There are over 170,000 people in 53,000 graves at Highgate Cemetery. The Victorian burial ground was opened in 1839 within the north-western wooded area. Highgate was designed as one of the Magnificent Seven modern cemeteries designed around the outside of London. The cemetery is Grade I listed, has many interesting tombs, and is the final resting place of several notable people associated with science.

Highgate Cemetery is split into the East and West side. Visitors can either do a self-guided tour or book a guided tour of the cemetery. However, the iconic catacombs can only be visited by guided tour. There are magnificent views of London from the Victorian terraced catacombs, considered the earliest surviving asphalted building in England. Other famous landmarks include the Egyptian Avenue and the Circle of Lebanon with its ancient cedar tree.

Scientists Buried in Highgate Cemetery

Notable scientists buried in Highgate Cemetery include the following people.

East Cemetery

Sir Thomas Lauder Brunton (1844–1916).
Scottish physician and pioneer of angina drug treatment.

William Clifford (1845–79).
Mathematician and philosopher who introduced geometric algebra.

Highgate Cemetery. (Packshot via Adobe)

Sir William Herdman (1858–1924). Scottish marine zoologist and oceanographer.

David Kirkaldy (1820–97). Scottish Engineer and pioneer of materials testing.

West Cemetery

Jacob Bronowski (1908–74). Scientist and broadcaster known for *The Ascent of Man* TV documentary series.

Michael Faraday (1791–1867). Natural philosopher and scientific adviser.

Robert Gardiner Hill (1811–78). A surgeon specialising in the treatment of the insane.

Frederick Pavy (1829–1911). Physician and physiologist who introduced the blood sugar test for diabetes.

HEATH ROBINSON'S PINNER

William Heath Robinson was born in North London in 1872 and worked as an artist and illustrator. He trained at Islington School of Art and wanted to work as a landscape painter but needed a more reliable income source. Heath Robinson illustrated books and became known for his creative designs. Of all his work, he is best known for his drawings and designs of weird gadgets and strange mechanical devices. Heath Robinson's name entered the everyday language as early as 1912 and is still used to describe bizarre contraptions of the type that ended up in his illustrations. Examples include rickety machines that somehow keep going with continual tinkering and adjustments. At Bletchley Park during the Second World War, the Wrens named one of the codebreaking machines 'The Heath Robinson'.

Heath Robinson moved to Hatch End near Pinner with his family in 1908 and later moved to Pinner itself, where he lived at 75 Moss Avenue.

Queen's Arms, Pinner (Author)

This is a private home, but a blue plaque on the wall commemorates its famous resident. Heath Robinson also enjoyed meeting friends and family at the Queen's Arms on the High Street in Pinner. Most of Heath Robinson's art, including the illustrations that established him as a humourist, was created during his time living in Pinner.

HEATH ROBINSON MUSEUM

Pinner Memorial Park, 50 West End Lane, Pinner, HA5 1AE
www.heathrobinsonmuseum.org

The Heath Robinson Museum is located in the delightful Pinner Memorial Park, close to Pinner Underground Station. It has displays of some of Heath Robinson's finest work, including his book illustrations and artwork. The museum follows his life from his early days at art school to his career as a leading illustrator when he had to look for more lucrative sources of income. There are both permanent and temporary exhibitions at the museum and an adjacent cafe overlooking the lake in the Memorial Park.

Heath Robinson Museum. (Author)

MUSEUMS WITH SCIENTIFIC, MEDICAL AND ENGINEERING HERITAGE

BENTLEY PRIORY MUSEUM

Mansion House Drive, Stanmore, HA7 3FB
https://bentleypriorymuseum.org.uk

Bentley Priory was pivotal as the headquarters of Fighter Command during the Battle of Britain. Today, it is a museum telling the story of the Battle of Britain through three engaging exhibits. Air Chief Marshal Sir Hugh Dowding was Commander in Chief during the Battle of Britain. The exhibition, The One, shows how his leadership was effective. It also includes the implementation of the world's first integrated air defence system, known as the Dowding System. The Few tells the story of the aircrew who fought in the Battle of Britain, and The Many is about the people on the ground. Visitors can learn how technology was used to protect Britain in the Second World War. The work of Sir Archibald McIndoe, who pioneered plastic surgery in injured pilots, is also recognised.

FREUD MUSEUM

20 Maresfield Gardens, London, NW3 5SX
www.freud.org.uk

Sigmund Freud is the founder of psychoanalysis, a concept that explains how the mind operates and a way of helping people in mental distress. He was born in 1856 in Freiburg in what was Austria and spent most of his life in Vienna. When the Nazis invaded Austria in 1938, Freud and his family moved to London to escape persecution as Jews. Freud died at his home in North London in 1939, and today the building at 20 Maresfield Gardens is a museum.

The house contains information and exhibits on the story of Freud and how psychoanalysis was developed. Freud's famous consulting couch is on display

Freud Museum. (Author)

along with personal items, statues such as a figure of Eros, and photos. There is also a significant display of the work of his daughter, Anna Freud, and her theory on child development. A model of her iconic toddler hut is on display. The garden is gorgeous, and somewhere the Freuds relaxed when living in London.

Islington Museum. (Author)

ISLINGTON MUSEUM

245 St John St, Islington, London, EC1V 4NB
www.islington.gov.uk/libraries-arts-and-heritage/heritage/islington-museum

Islington Museum might be small, but it is packed with local history. It is most well-known for its bust of Lenin, a one-time resident of Clerkenwell. However, there are several displays relating to science and health. Campaigning for disability rights began in Islington, and the museum displays the work. The efforts of Finsbury Health Centre is celebrated with its work on improving access to care. Another scientific achievement is an exhibit about Marie Stopes, who began the first British family planning clinic in nearby Holloway in 1921.

KEATS HOUSE MUSEUM

10 Keats Grove, Hampstead, NW3 2RR
www.londonshh.org/houses/keats-house-museum.html

John Keats is most famous for his poetry. Classic poems like 'To Autumn' and 'Ode to a Nightingale' are among the most famous in English literature. However, Keats was also apprenticed as a barber-surgeon to Thomas Hammond in 1810 following the death of his mother. Five years later, he registered as a medical student at Guy's Hospital and was soon employed as a dresser to the surgeons. Writing took up much of his time but he created some of the best romantic poetry of the age. Keats died of tuberculosis in 1821, aged just 25. The Keats House Museum in London contains medical objects associated with his work in the profession.

RAF MUSEUM, HENDON

Grahame Park Way, London, NW9 5LL
www.rafmuseum.org.uk/london

The RAF Museum in Hendon celebrates the work of the Royal Air Force, but it goes much further with its innovative displays and exhibitions. Learn how propellers work and take a journey in a flight simulator. The museum is split into three main exhibits portraying 100 years of the RAF, the current air force, and technology of the future. Visitors can also explore hangars and see how Bomber Command operated.

The RAF Museum offers opportunities to participate in behind-the-scenes tours where visitors can get up close to aircraft and learn how they operate. From a scientific perspective, there is the opportunity to learn about jet propulsion and why some material is better for building aircraft than others. In addition, the museum has a Typhoon simulator, a 4D Red Arrows experience, and an opportunity to find out how a Harrier takes off vertically.

VESTRY HOUSE MUSEUM

Vestry Road, Walthamstow, London, E17 9NH
https://vestryhousemuseum.org.uk

Vestry House has had a remarkable past. Dating from 1730, it was initially built as the parish workhouse and later became a police station, an armoury, and a private home. Today it houses a remarkable collection of items depicting the history of Waltham Forest. The museum includes displays relating to local manufacturers, including Ensign, who made lenses and cameras. There is also a display of early plastic products from the British Xylonite Company. Of particular scientific and engineering interest is the Bremer Car. Frederick Bremer built what is believed to be the world's first internal combustion engine car at his garage on Connaught Street in Walthamstow in 1892. It was built for personal use, so is unique.

Vestry House also has a good cafe and garden with a nature trail for children.

Vestry House Museum. (Author)

RESTAURANTS, PUBS AND CAFES WITH A SCIENTIFIC CONNECTION

CHIN CHIN LABS

Camden Market
https://chinchinicecream.com

Liquid nitrogen is usually used in science and medicine; however, a cafe in Camden uses it to make ice cream. Chin Chin Labs has a range of enticing flavours, and the ice cream is made on the premises as the liquid nitrogen causes the process to be much faster than conventional ice cream making. The colder temperature also enhances the flavour. It's a tiny cafe, but it's fun to watch your chosen ice cream being created in front of your eyes.

Chin Chin Labs in Camden. Making ice cream with liquid nitrogen. (Author)

THE ENGINEER

65 Gloucester Avenue, London, Greater London, NW1 8JH
www.theengineerprimrosehill.co.uk

The Engineer is located in Primrose Hill, a short walk from Camden Lock and Regent's Canal. It is a popular pub named after one of the most outstanding engineers, Isambard Kingdom Brunel.

The Engineer Pub. (Author)

ENGINEERING EXPERTISE

MARKFIELD BEAM ENGINE AND MUSEUM

Markfield Park, Tottenham, London, N15 4RB
www.mbeam.org

Today, Markfield Park is a recreation area, but it was once home to the Tottenham Sewage Works, providing an essential public health facility from the 1850s to 1964. The remarkable Markfield Beam Engine remains and is lovingly maintained by volunteers. Housed in a Grade II listed engine hall, the steam-powered beam engine was built in 1888. The engine is believed to be the last one built by Wood Brothers and is a rotary beam engine. It is also the only surviving eight-column engine in situ.

The museum has a small display about the engine and the role of sewage treatment in public health. In addition, volunteers hold steam-up sessions regularly where visitors can see the machine working, learn about its history from the team, and appreciate this beautiful Victorian engine in all its glory.

Markfield Beam Engine Hall. (Author)

DAYS OUT FROM LONDON

Several places with a rich scientific heritage are within an hour or two of London, making them ideal for a day visit or weekend break from the capital. They can easily be reached by public transport and include museums, places of scientific discovery, and more.

BLETCHLEY PARK MUSEUM AND THE NATIONAL MUSEUM OF COMPUTING

Bletchley Park, Sherwood Drive, Bletchley, MK3 6EB
https://bletchleypark.org.uk

Until recently, few people had heard of Bletchley Park and its role in the Second World War as it was shrouded in secrecy. During the war, Bletchley Park was the centre of codebreaking and worked on ways to outwit the enemy. Mathematicians, linguists, and scientists all worked on deciphering codes, breaking the Enigma Code, and even developing the first computers. However, due to the secrecy surrounding Bletchley Park, many people who worked on critical projects and made significant scientific breakthroughs were not recognised for their work.

Tips for Visiting

Bletchley Park and the National Computer Museum are on the same campus but operated separately. If you plan to visit both sites, organise your visit to coincide with opening times as they differ. You need at least half a day (sometimes longer) to see Bletchley Park. Visits must be pre-booked with an entry time via the websites. Special events, such as the afternoon tea in the original codebreakers' dining room, must be pre-booked, and for a good reason. They usually sell out.

Audio guides are available.

Getting There

Trains leave London Euston for Bletchley, arriving in under an hour. Bletchley Park is a short walk away. Turn right out of Bletchley Station onto Sherwood Drive,

Bletchley Park. (Lance Bellers via Adobe)

cross the pedestrian crossing, and Bletchley Park is 100 yards ahead.

Highlights at Bletchley Park

Block C gives an excellent introduction to the work at Bletchley Park, explaining some of the processes used, the difference between a cypher and a code, and more. You can also pick up a headset for an audio tour here.

The Mansion. The Library and Commander Denniston's office have been recreated inside the mansion as they were in the Second World War. Sometimes, concerts are held in the ballroom. But, if you booked afternoon tea, it is served in the original codebreakers' dining room.

Huts 3 and 6 have been restored so visitors get an insight into what it was like to work as a codebreaker. Listen to the voices telling their stories of the secretive experience at Bletchley Park.

Hut 8 gives insights into where the groundbreaking work of decoding the German Enigma codes took place with displays of how the offices were set up.

Bombe machine indicator dials at Bletchley Park. (Peter Greenway via Adobe)

Hut 11 housed the Turing-Welchman Bombe machines where Wrens worked day and night and is now an interactive museum display where you can hear the voices of the women who worked there in the war.

Museum in Block B houses the largest display of Enigma machines worldwide. It also highlights Alan Turing's work and shows the transition of the Government Code & Cypher School to today's GCHQ.

D-Day Exhibition. This immersive exhibition shows how the codebreakers worked to plan D-Day and their impact on the invasion that ended the war.

The Intelligence Factory. A new exhibition gives insights into the personal stories of people who worked at Bletchley Park, the Naval Plotting Room, and how the equipment was used, including an original Hollerith Machine.

HIGHLIGHTS OF THE NATIONAL MUSEUM OF COMPUTING

Block H, Bletchley Park, Sherwood Drive, Bletchley, MK3 6EB
www.tnmoc.org

The National Museum of Computing is much smaller than Bletchley Park but is packed with fascinating exhibits. It is home to the world's most extensive collection of working computers. A significant highlight is the staff, many of whom are volunteers. Their ability to explain the technicalities of the various displays in an engaging way to all levels of visitors, whether complete novices or diehard geeks, is remarkable.

Colossus digital computer, National Museum of Computing. (Steve Simmons UK via Adobe)

The Turing-Welchman Bombe

The museum has a working example of the famous Bombe that helped crack the Enigma Code and saved thousands of lives by shortening the war.

The Tunny Gallery

The Tunny Gallery demonstrates the entire codebreaking process from the German Lorenz encoding machines to final decryption in Bletchley Park.

The Heath Robinson

The machine, commonly called the Heath Robinson, was the device that inspired Colossus. It was used as an early attempt to automate codebreaking.

Colossus

The world's first electronic computer was built to decipher Lorenz-encrypted messages between Adolf Hitler and his senior staff. The machine was designed by Tommy Flowers at the Post Office Research Station in North-West London and delivered to Bletchley Park in 1943. It was the first electronic digital machine that could be programmed. However, the machine remained top secret for many years after the war ended.

Early Computers

The museum has an extensive collection of first-generation computers, including the Harwell Dekatron Computer (WITCH) and the first EDSAC computer.

DOWN HOUSE AND DOWNE VILLAGE, KENT

Charles Darwin is one of the most famous naturalists, having developed the Theory of Evolution. Although he made his iconic voyage on HMS *Beagle* and is famed for work in the Galapagos Islands, much of Darwin's experiments were conducted in his own backyard in Kent. Charles Darwin made Down House and the village of Downe his home for many years.

DOWN HOUSE

Luxted Rd, Downe, Orpington BR6 7JT

Tips for Visiting

Many people come to Down House and leave, but the village of Downe is worth exploring as there are several connections to the Darwin family. From Downe village, a public footpath leads to Down House across the fields and is safer than walking on the roads. Check the times of trains and connecting buses when planning a journey, as some are infrequent.

Getting There

The nearest train stations are Chelsfield, 3¾ miles; Orpington, 3¾ miles; and Bromley South, 5½ miles. TfL bus R8 from Orpington passes the entrance (except Sunday); TfL bus 146 from Bromley North and South terminates in Downe village, half a mile from the property.

Down House has a car park on the premises.

Down House, Charles Darwin's home. (Ivan via Adobe)

Highlights of Down House

When you visit Down House, you can literally walk in the footsteps of Charles Darwin. He lived here with his family for several years, and many of his experiments were conducted in the garden. The house displays Darwin's life on the first floor, with his study on the ground floor.

The Garden
Charles Darwin spent significant amounts of time in his garden, conducting experiments. The newly restored greenhouse and the flower beds outline how Darwin experimented with genetics using different varieties of plants, such as lupins. The Great House Meadow is where Darwin discovered the red clover's connection with cats and the interdependence between plant and animal species. Another area in the garden is the Wormstone. Darwin studied worms for over forty years, and his book on earthworms sold more copies than the *Evolution of Species*. He was able to show how crucial earthworms are to the fertility of the soil and how quickly they decompose vegetable matter. Darwin's Thinking Path is where he walked and considered aspects of nature and evolution.

First Floor
The first floor of Down House has an exciting exhibition of Darwin's life. There are displays of beetles and birds' eggs collected by Darwin and his voyage on the *Beagle*. The collection explains how Darwin's travels changed him and fuelled his interest in evolution. Darwin's bedroom is where the scientist died in 1882. From the windows, visitors can look out on the Great House Meadow, just as Darwin did when contemplating his theories.

Ground Floor
On the ground floor, Darwin's study is a significant highlight. He wrote *On the Origin of Species* in this room, which contains many objects from his time on HMS *Beagle*. All the items in the room are connected with Darwin somehow and have been restored to the 1870 arrangement.

HIGHLIGHTS OF DOWNE VILLAGE

In the village centre, the Queen's Head pub was a favourite of Darwin. It dates from 1559 and was named after Elizabeth I.

St Mary's Church, Downe

The Darwin family attended St Mary's Church, although Charles tended to go for a walk instead from 1849. Outside, the sundial on the church wall is Darwin's memorial. Some Darwin family members are buried in the churchyard,

St Mary's Church, Downe. (Author)

including Emma, Charles Darwin's wife. In addition, there are memorials to the Lubbock family in the church. Sir John Lubbock (1803–65) was an astronomer and the first Vice Chancellor of the University of London. His son, also Sir John Lubbock (1834–1913) was the first Lord Avebury, a scientist in archaeology and evolutionary theory and an educationalist. He was known as the 4th Baronet of Avebury from 1865 to 1900 and also founded bank holidays. The 4th Baronet was a close friend of Darwin and when the latter's theories were criticised from the pulpit here, Lubbock became upset and left for nearby Farnborough Church.

HISTORIC DOCKYARD, CHATHAM

The Historic Dockyard, Chatham, Kent, ME4 4TZ
https://thedockyard.co.uk

There are 400 years of heritage to explore at the Historic Dockyard in Chatham. Many attractions have science and engineering links and are within a short distance of central London.

Getting to Chatham

Trains leave St Pancras Station for Chatham Station twice an hour at peak times. There are also regular services from London Victoria and Charing Cross Stations. From Chatham station, there is either a thirty- to forty-minute walk or a bus that drops visitors at the entrance to the Dockyard complex on Western Avenue. Along the pavement opposite the station are metal plaques depicting the history of Chatham and ropemaking.

Tips for Visiting

General admission tickets for the Historic Dockyard at Chatham are valid for a year so visitors can return to see more of the attractions. The dockyard is used

HM Submarine Ocelot. (Scot Nicholl with permission from Historic Dockyard, Chatham)

frequently for filming, so some areas may have restrictions when visiting. Tickets work on a timed entry and should be pre-booked. *Call the Midwife* tours run separately to the general site and need to be booked in advance.

There is a large car park adjacent to the dockyard.

Scientific and Engineering Highlights

The Ropery

Rope has been made at Chatham for 400 years. The building is the only one of the four original Royal Navy Ropeworks to remain operational. The ¼ mile rope walk is housed in the largest brick building in Europe, where master ropemakers continue to work daily. Bicycles are used to get from one end to another at speed. There is also a display on the history of rope and its importance in the industry.

Rope at Chatham is made from a mixture of natural and synthetic fibres. Mathematics is also involved in the science of keeping ropes from unwinding. Jakob Bohr and Kasper Olsen from Denmark studied the number of times a strand in each rope is twisted. In traditional ropes, each

The Ropery at Historic Dockyard, Chatham. (Credit Louise Hubbard Photography with permission from Historic Dockyard, Chatham)

strand is twisted as much as possible in one direction. The twisted strands are wound together in a spiral shape known as a helix, which rotates in the opposite direction. When the twists and counter-twists interlock, it gives the rope strength so that when it is yanked or placed under pressure, it does not unwind. Bohr and Olsen discovered there is a maximum number of times that each strand can be twisted by plotting the rope's length against the number of twists. This phenomenon results in the zero twist point for the overall rope. A good strong rope will always be in the zero twist configuration.

Visitors can see the ropery and live demonstrations as part of the general admission ticket to the historic dockyard and buy rope made on the premises. There are also thirty-minute Ropery Experience tours where visitors can learn more about the art and science of ropemaking.

HMS Submarine *Ocelot* (1962)

Ocelot was built at Chatham in 1962 and is a diesel-powered submarine. The Oberon-class submarine was the

***Children looking through* Ocelot.** (Historic Dockyard, Chatham)

last naval warship built at Chatham and clocked up over 90,000 miles in service. It was mainly used in surveillance and on exercises around the world and was last in service in 1991.

Visitors can tour the submarine to see what life was like as a submariner and how they used physics to manage ventilation and depth underwater. The science of buoyancy and displacement of water is another critical element of submarine life. Look through the periscope, crawl through the hatches, and view the control room. Learn the difference between red and black light and how submarines use prisms. Tours are pre-timed. Due to the nature of the vessel, access is limited and is not suitable for people with limited mobility or who dislike enclosed spaces.

HMS *Cavalier*

In the Second World War, HMS *Cavalier* was brought into service as an emergency destroyer. Today, it serves as the National Destroyer Memorial to those who perished at sea. *Cavalier* is a CA-class destroyer and, after being launched in 1944, saw active service in the Western Approaches, British Pacific Fleet, and the Arctic.

The Admiralty commissioned the design to develop the use of electric welding in warship construction. *Cavalier* was one of the first ships to be made with a partially welded hull. Amidships remained riveted to ensure strength. The welding was successful because the new process enabled the ship to move faster. In addition, women could handle the welding more efficiently than traditional riveting, which freed men to enlist for the war effort.

Welding is critical in shipbuilding as poor-quality welds will cause structural collapse. To avoid faults like distortion and other imperfections, welders use controls such as temperature, voltage, and the speed of welding to ensure high-quality work.

Visitors can tour *Cavalier* within the dockyard.

HMS *Gannet*

HMS *Gannet* was built on the River Medway in 1878 as a sloop. It protected interests globally during the Victorian age and saw active service in the Mediterranean, Red Sea, and the South Pacific. *Gannet* was powered by steam and sail. The hull was constructed from teak with an iron frame. During service with the Royal Navy, the ship was involved in anti-piracy and slavery patrols. *Gannet* also worked on surveys including charting and hydrographics. It was also an accommodation ship to the nautical training school Mercury. The ship has been refitted to look as it did in 1888, and visitors can tour it as part of the dockyard experience.

No. 3 Covered Slip

When No. 3 Covered Slip was built at Chatham in 1838, it was the largest wide-span timber structure in Europe. The space is home to large objects from the dockyard and Royal Engineers Museum. Shipwright Sir Robert Seppings designed the incredible cantilever roof, which was at the peak of technological innovation when constructed.

Brunel's Sawmill Canal Lock (1812)

During the Napoleonic Wars, Chatham Dockyard needed vast supplies of timber to keep up with the demand for warships. As a result, the Navy built one of Britain's first steam-powered sawmills at Chatham. The site is in the central dockyard car park close to the entrance. The sawmill was designed by Sir Marc Isambard Brunel and connected by canal to the South Mast Pond, where timber was weathered. The canal was used to transport logs via a tunnel to the sawmill where wood was processed. The site was discovered by archaeologists in 2008.

Call the Midwife *official location tour.* (Historic Dockyard Chatham)

Call the Midwife Tours

The popular television series *Call the Midwife* is filmed at the Historic Dockyard. Tours including some of the popular filming locations and explanations of how buildings are disguised to look like London's East End are included. There is also an exhibition of medical equipment, props, and uniforms from the series for those on the guided tour only.

SOURCES AND FURTHER READING

City of London

Campbell, James W.P. (2007). *Building St Paul's*. Thames and Hudson.

Christopher, John (2012). *Wren's City of London Churches*. Amberley Publishing.

CNN News (2013) London Building Melts Car.

Desborough, Jane (2020). *Isaac Newton and the Royal Mint*. Science Museum.

Forshaw, Alec (2015). *Smithfield. Past, Present and Future*. Robert Hale.

Gardiner, Rena (2005). *The Story of St Bartholomew the Great*. Parish of St Bartholomew the Great.

Jardine, Lisa (2004). *The Curious Life of Robert Hooke. The man who measured London*. Harper Perennial.

Leotaud, Valentina Ruiz (2020). 'Origin of British Crown Jewels diamond revealed'. Mining.com

Lincoln, Margaret (2022). *London and the 17th Century. The making of the World's Greatest City*. Yale.

Mitchell, Piers D. (2011). *The Study of Anatomy in England From 1700 to the Early 20th Century*. Journal of Anatomy.

Riley-Smith, Jonathan (1999). *Hospitallers. The History of the Order of St John*. Bloomsbury.

Smith, Roff (2013). 'How sunlight reflected off a building can melt objects'. *National Geographic*.

Wise, Sarah (2005). *The Italian Boy. Murder and grave robbery in 1830s London*. Pimlico.

Wood, Caroline. 'Imitating life'. *The Biologist* 63(1), pp.12–16.

'Mathematical Tiles'. The Regency Town House, www.rth.org.uk

The West End and Westminster

Finnane, Heidi. 'Big Ben's Bong: The science behind its sound'. *Britain Magazine*.

Gillin, Edward J. (2017). *The Victorian Palace of Science. Scientific knowledge and the building of the Houses of Parliament*. Cambridge University Press.

Halliday, Stephen (2018). *The Great Stink of London. Sir Joseph Bazalgette and the Cleansing of the Victorian Metropolis*. The History Press.

Johnson, Stephen (2006). *The Ghost Map*. Riverside Books

Main, Jenny (2022). *Ethel Gordon Fenwick. Nursing reformer and first registered nurse*. Pen and Sword.

Moore, Wendy (2005). *The Knife Man. Blood, body-snatching, and the birth of modern surgery*. Bantam Books.

Moore, Wendy (2020). *Endell Street: The women who ran Britain's trailblazing military hospital*. Atlantic Books.

Pan, Song, et al. (2013). 'A review of the piston effect in subway stations'. *Advances in mechanical Engineering.*

Bloomsbury and Camden

Bradley, Simon (2007). *St Pancras Station.* Profile Books.
Fordham, Max, 'The physics of freezing at the Iranian Yakhchal', maxfordham.com
Greenwood, William. 'Seeing stars: astrolabes and the Islamic world'. British Museum
Joy, Jody. 'Lindow Man'. British Museum
'The Great Court of the British Museum', Archinomy, www.archinomy.com/case-studies/the-great-court-british-museum-london
Parkinson, Richard (2005). *The Rosetta Stone.* British Museum.

East London and Docklands

Garfield, Simon (2018). *Mauve: How one man invented a colour that changed the world.* Canongate.
www.livescience.com/21928-london-olympic-swimming-pool-tech.html
www.thomasrandallpage.com/Cody-Dock-Rolling-Bridge
Marshall, Geoff (2016). *London's Industrial Heritage.* The History Press.

South-East London

Agrawal, Roma (2021). *How Was That Built? The stories behind awesome structures.* Bloomsbury.
Burton, Anthony (2022). *The Brunels: Father and Son.* Pen and Sword
Christopher, John (2014). *Brunel in London.* Amberley Publishing.
Grogan, Suzie (2017). *Death, Disease, and Dissection. The Life of a Surgeon Apothecary 1750–1850.* Pen and Sword.
Halliday, Stephen (2009). *The Great Stink of London. Sir Joseph Bazalgette and the Cleansing of the Victorian Metropolis.* The History Press.
Jackson, Lee (2015). *Dirty Old London. The Victorian fight against filth.* Yale.
Kentley, Eric, Hulse, Robert, and Elton, Julia (2016). *The Brunel's Tunnel.* The Brunel Museum.
Rule, Christopher (2015). 'David Kirkaldy and his Testing and Experimenting Works'. Selia. www.howitworksdaily.com/how-do-pipe-organs-work
Yorke, Trevor (2018). *Victorian Pumping Stations.* Shire Books.

Greenwich

Lucas, Anya; Johns, Richard; Stewart, Sophie; Payne, Stephen (2019). *The Painted Hall. Sir James Thornhill's Masterpiece at Greenwich.* Merrell.
Jennings, Charles (1999). *Greenwich. The place where days begin and end.* Abacus.

Kensington, Chelsea and West London

Brown, Kevin (2000). *Alexander Fleming Laboratory Museum: A Guide.* St Mary's NHS Trust.

South-West London

Musgrave, Toby (2021). *The Multifarious Mr Banks. From Botany Bay to Kew, the Natural Historian Who Shaped the World*. Yale.

The Musical Museum. The Musical Museum at Kew Bridge Souvenir Guide.

Telscher, Kate (2020). *Palace of Palms. Tropical Dreams and the Making of Kew*. Picador.

North London

Ken Brereton (2013). *Tottenham Sewage Works and the Wood Bros. Beam Engine*. Markfield Beam Engine and Museum.

Days Out From London

Bohr Jakob, Olsen Kasper (2010). *The Ancient Art of Laying Rope*. Physics. Pop. Cornell University.

Costa, James T. (2017). *Darwin's Backyard. How small experiments led to a big theory*. Norton.

Erskine, Ralph, Smith, Michael (2011). *The Bletchley Park Code-Breakers*. Biteback Publishing.

Witze, Alexandra (2010). 'Physicists Untangle the Geometry of Rope'. *Science News*, 10 May 2010.

INDEX